T0205480

Additive Manufacturing Processes

Sanjay Kumar

Additive Manufacturing Processes

Sanjay Kumar
York University
North York, ON, Canada

ISBN 978-3-030-45091-5 ISBN 978-3-030-45089-2 (eBook)
https://doi.org/10.1007/978-3-030-45089-2

This Springer imprint is published by the registered company Springer Nature Switzerland AG
The registered company address is: Gewerbestrasse 11, 6330 Cham, Switzerland

Preface

Additive manufacturing (AM) is a relatively new manufacturing technique to make products. There is not a single book which exclusively deals with various AM processes, the present book is written to fulfil the gap. Emphasis is on communication.

The book is divided into 12 chapters. Chapter 1, named as Introduction defines an AM process and summarizes all AM processes. Chapter 2, named as Classification, attempts to classify all AM processes. There is yet no classification which takes into account all AM processes, and thus the given classification will provide the requisite classification. In order to address various categories of AM processes, new names such as powder bed process, photopolymer bed process, solid deposition process, liquid deposition process, air deposition process, ion deposition process, etc. are coined. Each of the chapters from Chapter 3 to 10 owes its name to the classification. Laser Powder Bed Fusion is the name of Chapter 3, Electron Beam Powder Bed Fusion is the name of Chapter 4 while Chapter 5 is Other Powder Bed Processes. Beam Based Solid Deposition Process is the name of Chapter 6 while Chapter 7 is Other Solid Deposition Processes. Chapter 8, named as Liquid Based Additive Layer Manufacturing consists of liquid bed process and liquid deposition process. Chapter 9, named as Air and Ion Deposition Processes deals with air deposition process and ion deposition process while Chapter 10 is Additive Non-Layer Manufacturing. Sheet based processes is placed in Chapter 11. There are some possibilities for future process development, which are given in the last chapter – Chapter 12.

I hope it will be useful for students, instructors, researchers and business professionals alike.

Sanjay Kumar

Abbreviations

AFSD	Additive Friction Stir Deposition
AJ	Aerosol Jetting
ALM	Additive Layer Manufacturing
AM	Additive Manufacturing
ANLM	Additive Non-Layer Manufacturing
ASTM	American Society for Testing and Materials
BAAM	Big Area Additive Manufacturing
BJ3DP	Binder Jet Three-Dimensional Printing
CAD	Computer-Aided Design
CEM	Composite Extrusion Modeling
CIB	Chemical-Induced Binding
CLF	Ceramic Laser Fusion
CLIP	Continuous Liquid Interface Production
CMT	Cold Metal Transfer
CNC	Computer Numerical Control
CSAM	Cold Spray Additive Manufacturing
DED	Directed Energy Deposition
DLP	Digital Light Processing
DMD	Digital Micromirror Device
EBAM	Electron Beam Additive Manufacturing
E-beam	Electron Beam
EBM	Electron Beam Melting
EPBF	Electron beam Powder Bed Fusion
ECAM	Electrochemical Additive Manufacturing
4 DP	Four Dimensional Printing
FDM	Fused Deposition Modeling
FGM	Functionally Graded Material
FLM	Fused Layer Modeling
FM	Full Melting
FPM	Fused Pellet Modeling
FSAM	Friction Stir Additive Manufacturing

FSBAM	Friction Surfacing Based Additive Manufacturing
FSP	Friction Stir Processing
FSW	Friction Stir Welding
GMAW	Gas Metal Arc Welding
GTAW	Gas Tungsten Arc Welding
HSS	High Speed Sintering
IJP	Ink Jet Printing
IM	Injection Moulding
LCM	Lithography based Ceramic Manufacturing
LDP	Liquid Deposition Process
LENS	Laser Engineered Net Shaping
LMHAM	Localized Microwave Heating based Additive Manufacturing
LOM	Laminated Object Manufacturing
LPBF	Laser Powder Bed Fusion
LPS	Liquid Phase Sintering
LSDP	Laser Solid Deposition Process
MAPS	Microheater Array Powder Sintering
MDDM	Micro Droplet Deposition Manufacturing
MJF	Multi Jet Fusion
PAD	Plasma Arc Additive Manufacturing
PBF	Powder Bed Fusion
PBNF	Powder Bed Non-Fusion
PBP	Powder Bed Process
PES	Projection based Electro-Stereolithography
PJ	Photopolymer Jetting
PME	Powder Melt Extrusion
PPBP	Photopolymer Bed Process
RFP	Rapid Freeze Prototyping
SBP	Sheet Based Process
SDP	Solid Deposition Process
SHS	Selective Heat Sintering
SIS	Selective Inhibition Sintering
SL	Stereolithography
SLM	Selective Laser Melting
SLS	Selective Laser Sintering
STL	Standard Triangle Language
3 DGP	3D Gel Printing
3 DP	Three Dimensional Printing
2 PP	Two-Photon Polymerization
T3DP	Thermoplastic 3D Printing
UC	Ultrasonic Consolidation
WAAM	Wire Arc Additive Manufacturing

Contents

Chapter 1
Introduction

Abstract Additive manufacturing (AM), a concept existing for the last 10,000 years, is defined and positioned among other manufacturing processes. The role of tools to separate processes is given. Its main difference from machining is highlighted, and the relation between its complexity and cost is visited. It is classified into two major categories: additive layer manufacturing and additive non-layer manufacturing. Various AM processes are summarized with few lines each, which will provide a quick glimpse into all AM processes. The roles of various processes in processing metals, ceramics, polymers, composites and functionally graded materials are succinctly mentioned.

Keywords Material · Layer · Tool · Machining · Complexity · Cost · Non-layer

1.1 Introduction

Additive manufacturing must mean a manufacturing that adds. But, if additive manufacturing means what it must mean then there is no need for a definition for additive manufacturing. This brings a question which types of additive manufacturing are called additive manufacturing in the present manufacturing practice and which types of additive manufacturing are excluded.

Additive process implies that materials are added with an aim either to improve an existing product or to make a new product. Examples of these processes are sintering, casting, injection moulding, stereolithography, ink jet printing, selective laser melting, directed energy deposition etc. These processes differ from subtractive process because the latter does not add the material but removes the material. Examples of subtractive processes are drilling, boring, CNC machining, milling, sawing, electrical discharge machining, laser ablation, water jet cutting etc. An additive process differs from a deforming process as the latter does not add the material but deforms the material. Examples of deforming processes are deep drawing, stamping, incremental forming, bending, forging etc. (Button 2014).

S. Kumar, *Additive Manufacturing Processes*,
https://doi.org/10.1007/978-3-030-45089-2_1

Examples of additive processes can be divided into two groups. Examples of the first group are sintering, casting, injection moulding, while examples of the second group are stereolithography (Hafkamp et al. 2017), ink jet printing (Derby 2015), selective laser sintering (Kumar 2003), electron beam melting (Cordero et al. 2017), directed energy deposition (Yan et al. 2018) etc. Examples of the first group are the first category of additive processes which requires design-specific tooling to make or develop a product while examples of the second group are the second category of additive processes which does not require design-specific tooling. This second category is known as additive manufacturing (AM).

It is the design-specific tooling which distinguishes AM from non-AM additive process. The role of tooling will be clear by taking examples of injection moulding (a non-AM additive process) and directed energy deposition (DED) (an AM process). In injection moulding (IM), a mould or a tool is required to make a part of one design, a number of parts can be made but all having the same design; if a part of another design is required, then another mould specific to that design is required – it implies that for making parts of a number of designs, a number of moulds will be required. In DED, a part is made by depositing materials in a certain path, if a part of another design is to be made then materials are deposited in another path; for a change in design of part, materials will be deposited by the same nozzle on the same platform and no extra investment on hardware is required – it implies that a number of parts having different designs can be made without having to go through a need to arrange a number of design-specific tools as happened in case of IM. This is how DED (an AM process) differs fundamentally from IM (a non-AM additive process).

It may be argued that some AM processes also require tools – why this requirement of tools (in AM) is not same as that requirement of tools (in non-AM additive). For example, additive friction stir deposition (AFSD) (Yu et al. 2018), an AM process, requires tool to accomplish the very process. In AFSD, if a design is very big, the tool will be worn out before it makes a complete part, many tools may be required to complete the same part pertaining to a single design; if the design is very small, the same tool will turn out to be oversized for that small design and will be unable to make a part as per the design without destroying the small features of the part, therefore a small tool is required in case of a small design. Thus, AFSD, an AM process, is not immune from the requirement of design-specific tools. This brings a question – how the requirement of design-specific tools in AFSD is different from the requirement of design-specific tools in IM. In IM, the design of the part is same as the design of the cavity of the tool or the shape of the part is same as the shape of the cavity of the tool, while in AFSD the design of the part need not be similar to either positive or negative replica of the tool. In AFSD, a number of tools are required so that fabrication as per any design will be completed with the help of either one tool or more tools, while in IM, a number of tools are required so that fabrication as per a particular design will be completed with the help of a particular tool.

1.2 Definition of a Process and Additive Manufacturing

Oxford dictionary gives definition of a process as 'a series of actions or steps taken in order to achieve a particular end'. AM processes are thus combination of various steps in sequence. Such steps are: what are the materials, how they are brought, how they are converted. For example, three steps in powder bed fusion (PBF) which define the process PBF are: materials are in the form of a powder, they are placed in the form of a bed, they are converted by fusion. Processes which come under PBF follow the same steps but differ in the source responsible for fusing them, for example, electron beam melting (EBM) and selective laser melting (SLM) are two PBF which differ because they are fused by electron beam and laser beam, respectively. Selective laser sintering (SLS) and SLM are two PBF which are fused by same type of source, that is laser beam, but they differ because both use the laser beam for different purposes; the former uses it for partial melting while the latter uses it for full melting. All these three PBF, that is SLS, SLM and EBM, differ in the last step of process steps; there are not notable examples of processes which differ in the first or the second step. It will not be surprising that in future PBF will emerge which will differ in the first step such as polymer or metal or ceramic or composite based or differ in the second step such as porous bed, curved bed, two beds etc.

A process is executed in a machine. Some machines execute more than one process while some processes can be executed in more than one machine. For example, an SLM machine can execute both processes, SLM as well as SLS; by decreasing laser power and preventing full melting, SLS can be achieved. Similarly, an SLS machine can execute both processes, SLS as well as SLM; by choosing low melting point materials such as tin or bronze and facilitating full melting SLM can be achieved. Thus a process (SLS or SLM) can be executed in two machines (SLS and SLM).

Most of the AM machines are process-specific and therefore these machines execute single AM process each. However, if other non-AM process can be included to understand a difference between process and machine, then these AM machines are found to execute more than one process. For example, a directed energy deposition (DED) machine can not only make a part, it can also perform laser cladding on another part, and thus execute two processes – DED and laser cladding; it is no wonder because DED is developed from laser cladding or DED is an extension of laser cladding. Most of the AM processes have similar humble beginning, and the related AM machine can execute two processes – one their initial process and other their existing process. For example, wire arc additive manufacturing (WAAM) is developed from arc welding, thus, a WAAM machine can execute two processes – arc welding (initial process) and WAAM (existing process). Examples of these AM processes are: ink jet printing is developed from paper printing, wire-electron beam additive manufacturing from electron welding, plasma arc additive manufacturing from plasma welding, friction stir additive manufacturing from friction stir welding, cold spray additive manufacturing from cold spray, electrochemical additive manufacturing from electro-printing etc.

1.3 Steps in Additive Layer Manufacturing

In additive layer manufacturing, a final part is visualized as an assembly of layers and therefore effort is done to make or to have layers and then assemble them. When a part needs to be fabricated then its Computer Aided Design (CAD) model is made which needs to be transformed in the form of layers; as shown in Fig. 1.1a, a cylindrical CAD model is transformed as an assembly of layers. Before a CAD model is transformed into layers (or is sliced), the models need to be converted into an assembly of many small interconnected triangles; this process of conversion is called tessellation which generates STL (standard triangle language or stereolithography) files. Triangles will not be able to exactly match the boundaries of the model if the boundary is not a straight line; therefore, in case of a model having curvature there remains a gap between the boundary of a CAD model and the boundary of an STL model. In order to slice this STL model, a horizontal plane intersects with the model at various points and collects the data at those points; the separation between two consecutive points is equal to the desired layer thickness (Roschli et al. 2019). These data determine the tool path – information is collected by machine in the

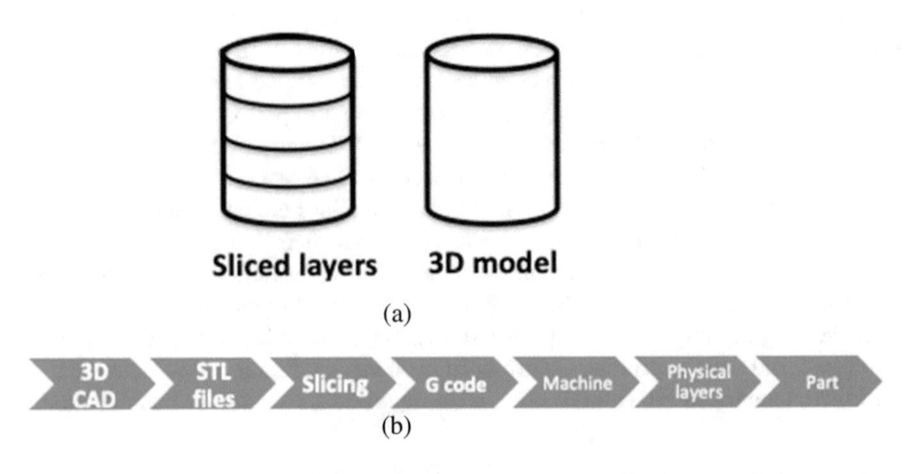

Fig. 1.1 Conversion of a CAD model into a part: (**a**) slicing of a cylindrical model into layers, (**b**) sequence of transformation

form of G-code – information required by the machine to use its tool accordingly and convert these data (virtual layers) into physical layers. The sequence of steps is given in Fig. 1.1b. The lowest layer is first fabricated followed by fabrication of successive layers over it; the fabrication of one layer after another continues till the highest or topmost layer is fabricated.

1.4 Additive Layer Manufacturing and Additive Manufacturing

Almost all AM processes are additive layer manufacturing (ALM), layerwise manufacturing is a synonym for AM and has become a norm for AM. It is no wonder that ASTM defines AM as ALM (ASTM 2012). Layer implies a horizontal layer while the build direction is vertical, all state-of-the-art AM machines follow the norm of horizontal layer-vertical build direction. There has been some attempt to make vertical layers and horizontal build direction; this is especially in the case of sheet metal manufacturing where big sheets need to be arranged vertically so that they can be aligned accurately. A rectangular sheet has a large cross-sectional area in comparison to the area of its side surface; if sheets are stacked horizontally, a large area needs to be aligned so that cavities and holes could coincide accurately. If they are stacked vertically, since vertical cross-section is far smaller than the horizontal cross-section of any rectangular sheet, only a small area of sheets needs to be aligned which is more convenient. Conversion of a CAD (Computer Aided Design) model into layers has been done for the sake of ease and convenience of manufacturing, conversion has not been done for the sake of not being in conflict with the basic principles of science and manufacturing. If the conversion is not done and the part is made without the help of design-specific tooling, there is nothing lost in the process that will hinder it to satisfy the basic concept of AM.

AM can be broadly classified into two processes: (1) additive layer manufacturing (ALM) and (2) additive non-layer manufacturing (ANLM) (Fig. 1.2). This is an asymmetrical classification where majority of AM processes go to the first category while a miniscule not-so-important processes go to the second category. This classification brings a question mark on the practical utility of this classification.

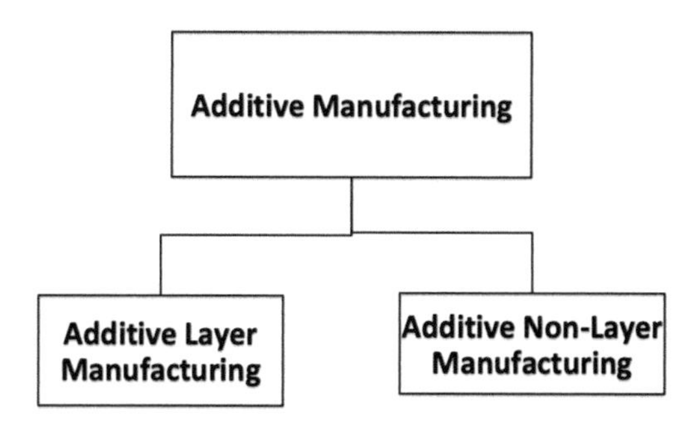

Fig. 1.2 Classification of AM on the basis of layer formation

Though, this classification may not have such practical utility, this classification is required not only to assert that AM is more than ALM but also to counteract an ever-increasing tendency that ALM is an ultimate manufacturing. AM (ALM) is considered an ultimate manufacturing solution for making complex parts, what if AM (ALM) fails. The classification is meant to emphasize that the failure of ALM is not a failure of AM because AM consists of ALM plus something (ANLM). Failure of ALM could not be a failure of AM because complete AM has not been tried. It does not imply that if complete AM will be tried then there will always be a solution for manufacturing, it only implies that the solution of a failure of ALM can still be found if complete AM (ALM plus ANLM) will be tried. In other words, failure of AM (ALM) is not a failure of AM (ANLM); when AM (ALM) fails, then AM (ANLM) can have the potential to provide an alternative path (given in Chaps. 9 and 10).

Examples of ALM are powder bed fusion (Grasso and Colosimo 2017), fused deposition modeling (Masood 2014) etc. while examples of ANLM are CNC accumulation (Chen et al. 2011), two photon polymerization (Nguyen and Narayan 2017) etc.

1.5 Oldest Evidence of Additive Layer Manufacturing

The earliest evidence of ALM happened 10,000 years ago when homes were first built brick by brick (Niroumand et al. 2013; Gallet et al. 2006) not different from layer by layer. When civilizations were borne out from bricks, when human beings laid the foundation of a nation. They were not only laying the foundation of a nation, they were also inadvertently laying the foundation of a notion of layer by layer – called additive layer manufacturing in modern times. Civilization is not there, the nation was destroyed, time passed by. Remaining structures started to fade. But ruins survived, it survived numerous onslaughts of thousand years and it was there to survive thunder of invading armies. In the ruins of that civilization, bricks are still there to say a saga. Once a magnificent wall exhibiting their art and glory is giving a glimpse that bricks were arranged layer by layer; ruins are still showing that walls were not monolithic structures but layered structures (comprising bricks of size ratio of $1 \times 2 \times 4$) (Dutt et al. 2019) – a type which modern manufacturing is so much proud of. Those who laid the foundation were the original inventers and innovators of ALM. They are not here to claim. They are not here to show their patents, there were no such patents, there was no such concept of a patent. The oldest evidence of AM as per patent is in the year 1890 (Bourell and Beaman 2003; Thompson et al. 2016). What if they appear and ask their concept back – ALM though having modern automation, electronics and computers at its disposal will not survive for a fraction of second without their concept. It is irony that the oldest evidence of the concept of ALM is in the field of construction, but it is only in recent times that the field has again seen the application of ALM (Camacho et al. 2018).

1.6 Comparison Between Machining and Additive Manufacturing

1.6.1 Approach

Machining is a big-small approach where a big block becomes a small product while AM is a small-big approach where small blocks become a big product. Thus, there are two different approaches. There are two separate approaches. This brings a question whether these two different approaches will still be different if these approaches will be observed through movements of tools in two approaches. Machining can be a top-down approach, while AM can be a bottom-up approach – these approaches can be due to the direction of tool movement – tool moves from top to down in machining while it moves from bottom to up in AM (Fig. 1.3). In machining, an object is made by machining a block from top surface. The object is continuously carved out from the block which results the object in continuously getting visible from the top surface to its down surface with the progress of machining. This is progressively shown from Fig. 1.3a–c through Fig. 1.3b where a block is cut by a machining tool. This definition does not imply that the machining has limitations and it cannot be performed from a side surface or a bottom surface, this only implies a general practice in machining. If machining is performed from the side surface or the bottom surface, then these surfaces will be termed as top surface. Top-down approach implies that the tool will go towards the material (block) so that it can machine while bottom-up approach implies that the tool will go away from the material (substrate or platform) so that it can have free space to build (Fig. 1.3).

In AM, build progresses from bottom surface or the first layer and reaches to the last layer or top surface. The bottom-up approach tells that it starts from the first layer, adds continuously layers and finishes at the last layer. If the orientation of a CAD model will change by 180° so that the top surface becomes bottom surface and vice versa, the approach does not change – it again needs to start from the first layer

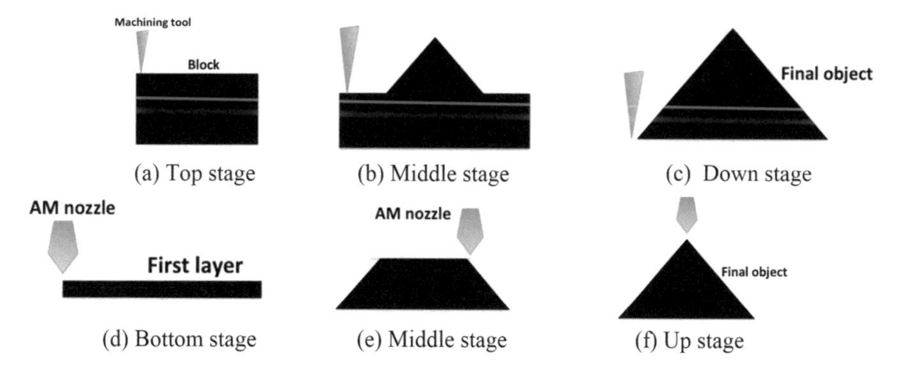

Fig. 1.3 Top-down approach in machining: (a) starting point, (b) middle stage, (c) final stage. Bottom-up approach in AM: (d) starting point, (e) middle stage, (f) final stage

which will then become the bottom surface of the build. If the orientation of the AM system changes by 180° (e. g. inverse stereolithography) so that the approach may look similar to top-down approach of the machining, but the approach will remain unchanged – it again needs to start from the first layer which will then become the bottom surface of the build. In summary, irrespective of the orientation of the final object, irrespective of the orientation of the system, the build starts from nothing, adds building blocks until it gets everything in the form of a final object – in bottom-up approach, fabrication happens through addition. In machining, the build starts from everything, removes blocks until there is nothing left to be removed except a shaped block in the form of a final object – in top-down approach, fabrication happens through disintegration.

Figure 1.3d–f shows the first stage, middle stage and the last stage of the build; these figures show an AM nozzle which is not common to all AM processes but the approach (bottom-up) is common to all AM processes. For processes, which do not use an AM nozzle, direction of build replaces the direction of movement of tool. In all AM processes, build direction starts from a bottom surface, which justifies the name 'bottom-up' approach.

The top-down and bottom-up approach is about the tool movement and does not indicate the orientation of AM systems or final AM products formed (Hafkamp et al. 2017; Santoliquido et al. 2019). Different approaches for getting different orientations of AM products on a substrate are given in Chap. 8.

1.6.2 Generation of Waste

While making an object, machining creates chips and swarfs that are waste. AM does not create chips and swarfs because it does not make an object by machining. There may be a small component of machining in AM in the form of modification or post-processing, which will create chips and swarfs – an amount miniscule in comparison to that created in only machining. Thus, AM is better than machining (a machining process) because it creates small amount or no amount of chips (waste) which machining cannot even dream for. There is no reason to doubt about the supremacy of AM over machining if creation of chips is measured. But, there is a reason to doubt about this method for creating supremacy. AM does not intend to do machining and therefore, if AM is free from the demerit of machining in comparison to a process which does only machining then the comparison is not on equal footing. The comparison would have been on equal footing if AM were doing as much machining as a process which does only machining – it could have been great if even then chips created by AM were miniscule.

It does not imply that if a process makes a product without machining or without creating chips through machining then it is not an achievement. It also does not imply that creation of chips is not discouraged. It also does not mean that the component of machining in AM should be increased in order to have a fair method for comparison. But, it does imply that the method for comparison is undue, it is undue because it misleads – it hides the fact that though AM does not make more chips, it may or may not make a lot more waste, it may or may not waste a lot of energy, it may or may not waste a lot of resources.

For example, if a nylon pattern is made each by machining and powder bed fusion (PBF) then the making of pattern by PBF will not create any waste by machining but will create waste by degradation of all powders present in a PBF system. The degradation of materials by PBF will be more than the waste of materials in the form of chip in machining process. Using PBF system will need more energy than that required for running a CNC machine, and thus the fabrication of a pattern by PBF system will waste more energy. This example does not represent all examples in AM, but this example shows that AM is not better than machining in all examples.

1.6.3 Material Properties

Machining does not change mechanical properties except some possible changes on a surface due to tool-induced heat and, therefore, properties of material before and after machining remain the same. AM does not come with such a material block and material property; it comes with a feedstock, experimental parameters, noise – which all have influences on material properties. Machining can be interesting because material properties will not deteriorate due to lack of skill in machining operation, or machining cannot be interesting because material properties will never improve even by adopting the best practices in machining operations. AM can be interesting because material properties at various locations of a part can be changed by optimizing parameters which will fulfil different functional requirements – this advantage is not available in machining. AM cannot be interesting because material property is an aggregate of contributions provided by tiny particles or pores or voxels or tiny volumes and, if any tiny entity is not controlled, the property may vary – machining is free from this challenge. But, this challenge has given rise to opportunities; if addition of tiny volumes of materials is controlled during manufacturing process then properties can be predicted; if properties can be predicted then properties can be known without measuring them; if properties can be known beforehand then desired properties can be introduced at a design stage (Roach et al. 2018). Difference between machining and AM in the context of laser beam deposition is given in Chap. 12.

1.7 Why 'Complexity in Additive Manufacturing Is Free' Is a Myth

AM is able to make a part incorporating cooling channels, undercuts, overhangs, cavities, features at various angles and dimensions; this ability of AM which results in complex components and products is unprecedented in the history of manufacturing and is one of the reasons for its growth and drawing worldwide attention. With the help of following example, the concept of complexity in AM is analysed from the perspective of cost.

If three rectangular parts of the same size are made from an AM system: (1) a rectangular part without any complexity, (2) a rectangular part having one cooling channel, this part is more complex than the previous part because it has a cooling channel, (3) a rectangular part having three cooling channels, this part is even more complex than the previous part because it has three cooling channels, but, this third part is not so complex that extra optimization of parameters is required. A set of parameters optimized for the first part works for all three parts. Thus, there is no extra cost for optimization of parameters. All three parts are of the same size so all will take same number of layers which will result in equal machine time – thus, the cost for machine time for all three parts are the same.

Thus, the complexity in three parts are increasing but the cost remains same; if complexity increases but the manufacturing cost does not increase, which is unprecedented, then there is nothing wrong to agree with the much-touted conclusion that complexity in AM is free (Friedman 2013; Hartmann 2015; Niaki et al. 2019).

If these three parts are made by conventional machining or casting, assuming that all parts can be made by going through many steps of machining or casting, then the cost will increase because the second part will not be manufactured using the same toolpath or same sequence of operations used for making the first part. Actually, the first part is easier and simpler to be made; in machining, it can be cut from an oversized block; in casting, a hollow rectangular mould which is easier to be machined will serve the purpose. Second part will require complexer mould. The third part will require more investment in casting mould than that required in case of the second part – it will result in more manufacturing cost for the third part than the second part. If the third part is going to be made by machining, then many steps of drilling will be required, these steps are certainly more than that required for the second part – it will result in more machining cost for the third part. Consequently, complexity in conventional manufacturing is not free, but it increases with an increase in complexity.

Above examples show as complexity increases from a simple rectangular part to a complex rectangular part, cost does not increase in AM while it increases in conventional manufacturing – thus, complexity in AM is free while complexity in conventional manufacturing is not free.

Above statements are logical. The statement ‘complexity in AM is free’ conveys a lot. But this statement hides a fact: if complexity in AM is free then what is the cost of simplicity in AM – simplicity in AM should be ultra-free. What is the cost of making a rectangular block in AM and what is the cost of making the same rectangular block in machining. The cost of making a rectangular block in AM is much more than that making the same in machining. If the size of a rectangular block will increase, the cost difference between AM and machining will again increase – with an increase in size of the part, the cost of making a simple part in AM will be much more than making a complex part in machining.

If complexity in AM is free, then it is because simplicity in AM is too much expensive. It does not imply that complex parts made in AM is not appreciable. It

Additive Manufacturing	Restaurant
Complexity = Free More complexity = Free Any complexity = Free Cost starting from = $5000	Coke = Free Coffee = Free Tea = Free Entry fee = $1000

Fig. 1.4 Highlighting the concept of complexity in AM

also does not imply that AM should focus on making simpler parts and try to prove that it can make as inexpensive parts as in machining. It only implies that the statement 'complexity in AM is free' is half information. Saying 'complexity in AM is free' is similar to saying 'drink in a restaurant is free' without mentioning the heavy entry fee of the restaurant (Fig. 1.4).

1.8 Summary of Additive Manufacturing Processes

Various processes are given briefly which are described in details in successive chapters. All major processes along with their most prevalent names are given; the abbreviation of the names is used for further categorization and explanation.

1. **Additive friction stir deposition (AFSD) :** Powder or rod is deposited and joined on a substrate using friction and stir caused by a tool (Yu et al. 2018).
2. **Aerosol jetting (AJ) :** Aerosol is deposited to make micro-scale objects (Goh et al. 2018; Johannes et al. 2018), also known as aerosol jet printing (Wilkinson et al. 2019).
3. **Binder jet three-dimensional printing (BJ3DP):** Binder is jetted on powder bed to make layers (Enneti et al. 2018). Also known as 3D printing (Kernan et al. 2007), binder jetting printing (Kunchala and Kappagantula 2018).
4. **Big area additive manufacturing (BAAM):** It is fused deposition modelling applied to make big parts. Instead of filaments, pellets are used (Roschli et al. 2019).
5. **Ceramic laser fusion (CLF):** Slurry is spread on a platform and is shaped by a laser beam (Tang 2002), also known as layerwise slurry deposition (Muhler et al. 2015). In one of its variants named selective laser gelling, sol of the slurry gets converted into gel by an applied laser beam (Liu and Liao 2010; Liu et al. 2013). In another variant named selective laser gasifying of frozen slurry, ice of frozen slurry bed is gasified by laser beam to make porous structures (Zhang et al. 2017).

(continued)

6. **Cold spray additive manufacturing (CSAM):** Material is projected at high speed on a platform to make a 3D object (Yin et al. 2018).
7. **Composite extrusion modeling (CEM):** It is extrusion based additive manufacturing in which composite powders are melted and extruded (Lieberwirth et al. 2017).
8. **Continuous liquid interface production (CLIP):** Parts from photopolymer resin are formed without layer-by-layer process (Janusziewicz et al. 2016).
9. **Digital light processing (DLP):** Selective solidification of photopolymer liquid is done by digital micromirror device (Salonitis 2014).
10. **Electrochemical additive manufacturing (ECAM):** Ions are deposited on an electrode to make 3D objects (Kamraj et al. 2016).
11. **Electron beam melting (EBM):** It is electron beam based powder bed fusion (EPBF). Powders are spread on a platform and are joined by an electron beam, also known as powder bed fusion (Korner 2016; Cordero et al. 2017; Gong et al. 2014).
12. **Electron beam additive manufacturing (EBAM):** Metal wire is deposited on a platform using an electron beam (Tarasov et al. 2019; Fox and Beuth 2013).
13. **Fused deposition modeling (FDM):** It is extrusion based additive manufacturing in which polymer based filaments are melted and extruded (Masood 2014). Also known as fused filament fabrication (Brenken et al. 2018), material extrusion process etc.
14. **Fused pellet modeling (FPM):** It is extrusion based additive manufacturing in which polymer based pellets are melted and extruded (Wang et al. 2016), also called fused layer modeling (Kumar et al. 2018).
15. **High speed sintering (HSS):** Each layer is heated with an infrared lamp after selectively jetting ink on the layer (Brown et al. 2018; Thomas et al. 2006). A variant of HSS with an option to deposit ink at boundary is named as multi-jet fusion (MJF) (Sillani et al. 2019).
16. **Ink jet printing (IJP):** Ink is deposited on a platform to make 3D objects from ink (Stringer and Derby 2009; Derby 2015).
17. **Laser engineered net shaping (LENS):** Powders are blown on a platform and are joined by a laser beam. Also known as direct metal deposition, blown powder technique, powder fed process, directed energy deposition DED (Yan et al. 2018), laser powder deposition (Vilar 2014). In one of its variants, when a single crystal is grown, the process is named as scanning laser epitaxy (Kirka et al. 2009).
18. **Lithography based ceramic manufacturing (LCM):** It is SL using photopolymer based ceramic slurry (Harrer et al. 2017; Schwarzer et al. 2017).
19. **Localized microwave heating based AM (LMHAM)**: Microwave energy supplied by microwave applicator acts as a heat source to sinter powder layer (Jerby et al. 2015).

(continued)

20. **Micro droplet deposition manufacturing (MDDM):** Metal droplets of low melting point metal are deposited by a nozzle to make 3D structures (Chao et al. 2012; Zuo et al. 2016).
21. **Microheater array powder sintering (MAPS)**: This is a variant of SLS in which an array of micro heater replaces laser beam as a heat source (Holt et al. 2018).
22. **Photopolymer jetting (PJ)**: Photopolymer is deposited on a platform to make 3D objects (Lanceros-Méndez and Costa 2018).
23. **Plasma arc additive manufacturing (PAD)**: Metal wire is melted by plasma beam. Also known as rapid plasma deposition (Feng et al. 2018).
24. **Powder melt extrusion (PME):** It is extrusion based additive manufacturing in which polymer powders are melted and extruded (Boyle et al. 2019).
25. **Rapid freeze prototyping (RFP):** Water drop is frozen to make a 3D object (Bryant et al. 2003), its variant is cryogenic prototyping (Pham et al. 2008) that uses a solution instead of water.
26. **Selective heat sintering (SHS):** It is powder bed fusion in which fusion is done by thermal printheads (Baumers et al. 2015).
27. **Selective inhibition sintering (SIS):** It is a variant of BJ3DP in which anti-binder ink instead of binder ink is selectively deposited. During post-processing in furnace, inked area is not joined while remaining area gets joined (Khoshnevis et al. 2014).
28. **Selective laser melting (SLM)**: Powders are spread on a platform and are joined by melting powders using a laser beam. Also known as direct metal laser sintering (Kumar 2014). In one of its variants named microwave-assisted selective laser melting, microwave energy is used to preheat the substrate which lets low laser power to melt ceramics (Buls et al. 2018).
29. **Selective laser sintering (SLS):** Powders are spread on a platform and are joined by a laser beam. This is also called laser sintering or powder bed fusion (Kumar 2003). In one of its variants named selective laser flash sintering, electric field is applied across powder bed to improve sintering rate (Hagen et al. 2018; Hagen et al. 2019).
30. **Stereolithography (SL):** Photopolymer is solidified by a scanning beam. Also known as vat photopolymerization (Hafkamp et al. 2017), scan based polymerization, micro-stereolithography and large area maskless photopolymerization (Rudraraju and Das 2009).
31. **Thermoplastic 3D printing (T3DP):** Hard particles mixed with thermoplastics are deposited to make a part (Scheithauer et al. 2017).
32. **3D gel printing (3DGP):** It is a slurry deposition process in which slurry is mixed with cross-linking polymers for binding (Ren et al. 2016).
33. **Two-photon polymerization (2PP):** Free-form object is formed by interaction of two photons in photopolymer (Nguyen and Narayan 2017; Wu et al. 2006).
34. **Wire arc additive manufacturing (WAAM):** Wire is melted by an arc to make a 3D object (Tabernero et al. 2018; Cunningham et al. 2018).

1.9 AM Processes for Fabricating Parts

Various AM processes used for making metallic, polymer, ceramic, composite parts and functionally graded materials are given below.

1.9.1 Metallic Parts

Processes which are used to make metallic parts directly are given along with brief notes. Here, 'directly' means no secondary process such as casting or infiltration is needed to make parts (Karapatis et al. 1998).

1. SLS: The process relies on partial melting. Metallic parts are formed by partially melting metal powders or by using a composite powder having a low melting point component or a metal powder coated with a binder. In order to get fully dense metallic parts, porous parts need to be infiltrated with lower melting point metals.
2. SLM: The process relies on full melting and is predominantly used to make complex parts.
3. MDDM: Low melting point metals are melted and deposited.
4. EBM: The process works in a vacuum environment and is especially suitable for metals such as aluminium or titanium which may degrade or burn in presence of oxygen.
5. BJ3DP: Metallic powders are joined using binder to make metallic parts but the part is not as strong as made by full melting of metals.
6. CLF: By using metallic slurry and oven treatment of CLF parts, a metallic part consisting of two metals can be formed.
7. LENS, EBAM, WAAM: These are exclusively used to make metallic parts.
8. CSAM: This process gives better part with low melting point metals.
9. ECAM: By depositing metallic ion on various shapes of cathodes, complex parts are formed.
10. CEM: By using pellet consisting of metals, a metal part after post-processing is obtained.
11. AJ: Used to make metallic electronic circuit.

1.9.2 Polymer Parts

Processes used to make polymer parts directly are given along with brief notes.

1. SLS: It is used to make complex parts from polymer powders.
2. HSS: Energy-efficient fabrication of plastic parts.

3. SHS: Using thermal printheads to make nylon parts.
4. SLM: By completely melting polymer, a term 'SLM of polymer' can be used.
5. BJ3DP: Polymer powders are joined using binder to make parts.
6. IJP: Polymer ink is used to make parts.
7. SL, DLP, PJ: These make parts with high resolution using photopolymer.
8. 2PP: It is used to make micron size parts having nanometre resolution from photopolymer.
9. FDM, BAAM: This is mainly used to make parts using filaments and pellets.

1.9.3 Ceramic Parts

Processes used for making ceramic parts directly are given along with brief notes.

1. SLS: Binder-coated ceramic powder is used to make parts.
2. SLM: Completely melting ceramic powders at high platform temperature has given crack-free parts.
3. BJ3DP: It is used to make complex parts.
4. CLF: Ceramic slurry is shaped by CLF and heat-treated to perform sintering.
5. IJP: Ceramic ink is widely used.
6. AJ: Ceramic electronic circuit and component are made.
7. SL, DLP, LCM: Used to make ceramic part by indirect process. Photopolymer containing ceramic is used to make parts.

1.9.4 Composite Parts

Processes used for making composite parts directly are given along with brief notes.

1. SLS: Composite powders are processed or infiltration is used to make parts.
2. SLM: Composite powder is melted.
3. BJ3DP: Either composite powder is processed or infiltration is used on porous BJ3DP parts.
4. CLF: Ceramic slurry is processed and sintered afterwards or infiltrated with other materials.
5. LENS: A mixture of metal and ceramic powder is used to make parts.
6. ECAM: Colloidal particles are deposited.
7. IJP: Composite ink is used.
8. T3DP: Thermoplastic mixed with composite is deposited.
9. FDM, CEM: Composite filament or pellet is used.
10. SL: Photopolymer containing ceramic is used.
11. PES: Another particle is projected on photopolymer layer.

1.9.5 Functionally Graded Materials

Processes used to make Functionally Graded Materials (FGM) (Zhang et al. 2019) directly are given along with brief notes.

1. SLS: By changing process parameters, a gradient of porosity is created giving rise to FGM.
2. SLM, EBM: By changing process parameters, porosity gradient is created or microstructure is changed.
3. LENS: By changing feed rate of two different powders, material composition is varied.
4. CSAM: By changing impact speed with a change in position, material composition is varied.
5. ECAM: By changing current, amount of deposition is changed giving rise to FGM.
6. IJP: Part density is constantly varied by changing deposition density per unit area.

References

ASTM F2792-12a (2012) Standard terminology for additive manufacturing technologies (withdrawn 2015). ASTM International, West Conschohocken

Baumers M, Tuck C, Hague R (2015) Selective heat sintering versus laser sintering: comparison of deposition rate, process energy consumption and cost performance. In: SFF proceedings, pp 109–121

Bourell DL, Beaman JJ (2003) Chronology and current processes for freeform fabrication. J Jpn Soc Powder Metall 50(11):981–991

Boyle BM, Xiong PT, Mensch TE et al (2019) 3D printing using powder melt extrusion. Addit Manuf 29:100811

Brenken B, Barocio E, Favaloro A et al (2018) Fused filament fabrication of fiber-reinforced polymers: a review. Addit Manuf 21:1–16

Brown R, Morgan CT, Majweski CE (2018) Not just nylon—improving the range of materials for high speed sintering. In: SFF proceedings, pp 1487–1498

Bryant FD, Sui G, Leu MC (2003) A study on effects of process parameters in rapid freeze prototyping. Rapid Prototyp J 9(1):19–23

Buls S, Vleugels J, Hooreweder BV (2018) Microwave assisted selective laser melting of technical ceramics. In: SFF proceedings, pp 2349–2357

Button ST (2014) Introduction to advanced forming technologies. Compr Mater Process 3:1–5. Elsevier

Camacho DD, Clayton P, O'Brien WJ et al (2018) Applications of additive manufacturing in the construction industry- a forward-looking view. Autom Constr 89:110–119

Chao Y, Qi L, Xiao Y et al (2012) Manufacturing of micro thin-walled metal parts by microdroplet deposition. J Mater Process Technol 212(2):484–491

Chen Y, Zhou C, Lao J (2011) A layerless additive manufacturing process based on CNC accumulation. Rapid Prototyp J 17(3):218–227

Cordero ZC, Meyer HM III, Nandwana P, Dehoff RR (2017) Powder bed charging during electron-beam additive manufacturing. Acta Mater 124:437–445

Cunningham CR, Flynn JM, Shokrani A et al (2018) Invited review article: strategies and processes for high quality wire arc additive manufacturing. Addit Manuf 22:672–686
Derby B (2015) Additive manufacturing of ceramic components by ink jet printing. Engineering 1(1):113–123
Dutt S, Gupta AK, Singh M et al (2019) Climate variability and evolution of the Indus civilization. Quat Int 507:15–23
Enneti RK, Prough KC, Wolfe TA et al (2018) Sintering of WC-12%Co processed by binder jet 3D printing (BJ3DP) technology. Int J Refract Met Hard Mater 71:28–35
Feng Y, Zhan B, He J, Wang K (2018) The double-wire feed and plasma arc additive manufacturing process for deposition in Cr-Ni stainless steel. J Mater Process Technol 259:206–215
Fox J, Beuth J (2013) Process mapping of transient melt pool response in wire feed e-beam additive manufacturing of Ti-6Al-4V. In: SFF proceedings, pp 675–683
Friedman T (2013) When complexity is free. The New York Times
Gallet Y, Genevey A, Goff ML et al (2006) Possible impact of the Earth's magnetic field on the history of ancient civilizations. Earth Planet Sci Lett 246(1–2):17–26
Goh GL, Agarwala S, Tan YJ, Yeong WY (2018) A low cost and flexible carbon nanotube pH sensor fabricated using aerosol jet technology for live cell applications. Sensors Actuators B Chem 260:227–235
Gong X, Anderson T, Chou K (2014) Review on powder-based electron beam additive manufacturing technology. Manuf Rev 1(2):1–12
Grasso M, Colosimo BM (2017) Process defects and insitu monitoring methods in metal powder bed fusion: a review. Meas Sci Technol 28:044005
Hafkamp T, Baars GV, Jager BD, Etman P (2017) A trade-off analysis of recoating methods for vat photopolymerization of ceramics. In: SFF proceedings, vol 28, pp 687–711
Hagen D, Kovar D, Beaman JJ (2018) Effects of electric field on selective laser sintering of yttria-stabilized zirconia ceramic powder. In: SFF symposium proceedings, pp 909–913
Hagen D, Chen A, Beaman JJ, Kovar D (2019) Selective laser flash sintering of yttria-stabilized zirconia. In: SFF symposium proceedings
Harrer W, Schwentenwein M, Lube T, Danzer R (2017) Fractography of zirconia-specimens made using additive manufacturing (LCM) technology. J Eur Ceram Soc 37:4331–4338
Hartmann D (2015) Complexity is free. Lulu Publishing Services
Holt N, Horn AV, Montazeri M, Zhou W (2018) Microheater array powder sintering: a novel additive manufacturing process. J Manuf Process 31:536–551
Janusziewicz R, Tumbleston JR, Quintanilla AL et al (2016) Layerless fabrication with continuous liquid interface production. PNAS 11(42):11703–11708
Jerby E, Meir Y, Salzberg A et al (2015) Incremental metal-powder solidification by localized microwave-heating and its potential for additive manufacturing. Addit Manuf 6:53–66
Johannes SJ, Keicher DM, Lavin JM et al (2018) Multimaterial aerosol jet printing of passive circuit elements. In: SFF symposium proceedings, pp 473–478
Kamraj A, Lewis S, Sundaram M (2016) Numerical study of localized electrochemical deposition for micro electrochemical additive manufacturing. Procedia CIRP 42:788–792
Karapatis NP, Van Griethuysen JPS, Glardon R (1998) Direct rapid tooling: a review of current research. Rapid Prototyp J 4(2):77–89
Kernan BD, Sachs EM, Oliveira MA, Cima MJ (2007) Three dimensional printing of tungsten carbide-10 wt % cobalt using a cobalt oxide precursor. Int J Refract Met Hard Mater 25:82–94
Khoshnevis B, Zhang J, Fateri M, Xiao Z (2014) Ceramics 3D printing by selective inhibition sintering. In: SFF proceedings, pp 163–169
Kirka M, Bansal R, Das S (2009) Recent progress on scanning laser epitaxy: a new technique for growing single crystal superalloys. In: SFF proceedings, pp 799–806
Korner C (2016) Additive manufacturing of metallic components by selective electron beam melting- a review. Int Mater Rev 61(5):361–377
Kumar S (2003) Selective laser sintering-a qualitative and objective approach. JOM 55(10):43–47
Kumar S (2014) Selective laser sintering/melting. Compr Mater Process 10:93–134

Kumar N, Jain PK, Tandon P, Pandey PM (2018) Investigation on the effects of process parameters in CNC assisted pellet based fused layer modeling process. J Manuf Process 35:428–436
Kunchala P, Kappagantula K (2018) 3D printing high density ceramics using binder jetting with nanoparticle densifiers. Mater Des 155:443–450
Lanceros-Méndez S, Costa CM (2018) Printed batteries: materials, technologies and applications. Wiley, Hoboken
Lieberwirth C, Harder A, Seitz H (2017) Extrusion based additive manufacturing. J Mech Eng Autom 7:79–83
Liu FH, Liao YS (2010) Fabrication of inner complex ceramic parts by selective laser gelling. J Eur Ceram Soc 30(16):3283–3289
Liu FH, Lee RT, Lin WH, Liao YS (2013) Selective laser sintering of bio-metal scaffold. Procedia CIRP 5:83–87
Masood SH (2014) Advances in fused deposition modeling. Compr Mater Process 10:69–91
Muhler T, Gomes C, Ascheri M et al (2015) Slurry-based powder beds for selective laser sintering of silicate ceramics. J Ceram Sci Technol 06(02):113–118
Nguyen AK, Narayan RJ (2017) Two-photon polymerization for biological applications. Mater Today 20(6):314–322
Niaki MK, Torabi SA, Nonino F (2019) Why manufacturers adopt additive manufacturing technologies: the role of sustainability. J Clean Prod 222:381–392
Niroumand H, Zain MFM, Jamil M, Niroumand S (2013) Earth architecture from ancient until today. Procedia Soc Behav Sci 89:222–225
Pham CB, Leong KF, Lim TC, Chian KS (2008) Rapid freeze prototyping technique in bio-plotters for tissue scaffold fabrication. Rapid Prototyp J 14(4):246–253
Ren X, Shao H, Lin T, Zheng H (2016) 3D gel-printing- an additive manufacturing method for producing complex shaped parts. Mater Des 101:80–87
Roach RA, Bishop JE, Johnson K et al (2018) Using additive manufacturing as a pathway to change the qualification paradigm. In: SFF symposium proceedings, pp 3–13
Roschli A, Gaul KT, Boulger AM et al (2019) Designing for big area additive manufacturing. Addit Manuf 25:275–285
Rudraraju A, Das S (2009) Digital date processing strategies for large area maskless photopolymerization. In: SFF symposium proceedings, pp 299–307
Salonitis K (2014) Stereolithography. Compr Mat Process 10:19–67. Elsevier
Santoliquido O, Colombo P, Ortona A (2019) Additive manufacturing of ceramic components by digital light processing: a comparison between the "bottom-up" and the "top-down" approaches. J Eur Ceram Soc 39(6):2140–2148
Scheithauer U, Potschke J, Weingarten S et al (2017) Droplet-based additive manufacturing of hard metal components by thermoplastic 3D printing (T3DP). J Ceram Sci Technol 8(1):155–160
Schwarzer E, Götz M, Markova D et al (2017) Lithography-based ceramic manufacturing (LCM) – viscosity and cleaning as two quality influencing steps in the process chain of printing green parts. J Eur Ceram Soc 37(16):5329–5338
Sillani F, Kleijnen RG, Vetterli M et al (2019) Selective laser sintering and multi jet fusion: process-induced modification of the raw materials and analyses of parts performance. Addit Manuf 27:32–41
Stringer J, Derby B (2009) Limits to feature size and resolution in ink-jet printing. J Eur Ceram Soc 29:913–918
Tabernero I, Paskual A, Alvarez P, Suarez A (2018) Study on arc welding processes for high deposition rate additive manufacturing. Procedia CIRP 68:358–362
Tang HH (2002) Direct laser fusing to form ceramic parts. Rapid Prototyp J 8(5):284–289
Tarasov SY, Filippov AV, Shamarin NN et al (2019) Microstructural evolution and chemical corrosion of electron beam wire-feed additively manufactured AISI 304 stainless steel. J Alloys Compd 803:364–370
Thomas HR, Hopkinson N, Erasenthiran P (2006) High speed sintering- continuing research into a new rapid manufacturing process. In: SFF proceedings, pp 682–691

Thompson MK, Moroni G, Vaneker T et al (2016) Design for additive manufacturing: trends, opportunities, considerations, and constraints. CIRP Ann 65(2):737–760
Vilar R (2014) Laser powder deposition. Compr Mater Process 10:163–216. Elsevier Ltd
Wang Z, Liu R, Sparks T, Liou F (2016) Large scale deposition system by an industrial robot (I): design of fused pellet modeling system and extrusion process analysis. 3D Print Addit Manuf 3(1):39–47
Wilkinson NJ, Smith MAA, Kay RW et al (2019) A review of aerosol jet printing – a non-traditional hybrid process for micro-manufacturing. Int J Adv Manuf Technol 105:1–21
Wu S, Serbin J, Gu M (2006) Two-photon polymerization for three-dimensional micro-fabrication. J Photochem Photobiol A Chem 181:1–11
Yan Z, Liu W, Tang Z et al (2018) Review on thermal analysis in laser-based additive manufacturing. Opt Laser Technol 106:427–441
Yin S, Cavaliere P, Aldwell B et al (2018) Cold spray additive manufacturing and repair: fundamentals and applications. Addit Manuf 21:628–650
Yu HZ, Jones ME, Brady GW et al (2018) Non-beam-based metal additive manufacturing enabled by additive friction stir deposition. Scr Mater 153:122–130
Zhang G, Chen H, Zhou H (2017) Additive manufacturing of green ceramic by selective laser gasifying of frozen slurry. J Eur Ceram Soc 37(7):2679–2684
Zhang C, Chen F, Huang Z et al (2019) Additive manufacturing of functionally graded materials: a review. Mater Sci Eng A 764:138209
Zuo H, Li H, Qi L, Zhong S (2016) Influence of interfacial bonding between metal droplets on tensile properties of 7075 Aluminum billets by additive manufacturing technique. J Mater Sci Technol 32(5):485–488

Chapter 2
Classification

Abstract In order to classify additive manufacturing (AM) processes, it is checked why they are different from each other. Their differences in terms of materials, energy sources, types of feedstocks used and conveyance of feedstocks are studied and attempts are done to classify on the basis of these differences. AM processes, whether they are additive layer manufacturing type or additive non-layer manufacturing type, are each classified into three types: material bed process, material deposition process and motionless material process. This classification allowed to accommodate all existing AM processes. Besides, this classification provides plenty of space to accommodate future AM processes.

Keywords Material · Feedstock · Layer · Non-layer · Energy sources

2.1 Introduction

There are many additive manufacturing (AM) processes available (given in Chap. 1); there are different names for the same process. Majority of the processes with most prevalent names are given in Table 2.1. These processes are similar in many respects and different in many respects – it leads them to be grouped in different ways. These groupings will lead them to be classified. This chapter provides various ways of grouping and find more suitable methods to classify. The search is to find a classification which will not only be able to accommodate all AM processes but also be broad enough to accommodate AM processes which are yet to be invented.

2.2 Difference in Additive Manufacturing Processes

There are many AM processes, they are different, their differences arise from the following sources.

S. Kumar, *Additive Manufacturing Processes*,
https://doi.org/10.1007/978-3-030-45089-2_2

Table 2.1 List of majority of AM processes

Process	Acronym
Additive friction stir deposition	AFSD (Yu et al. 2018)
Aerosol Jetting	AJ (Goh et al. 2018; Johannes et al. 2018)
Big area additive manufacturing	BAAM (Roschli et al. 2019)
Binder jet three dimensional Printing	BJ3DP (Enneti et al. 2018; Kernan et al. 2007)
Ceramic laser fusion	CLF (Tang 2002)
Cold spray additive manufacturing	CSAM (Yin et al. 2018)
Continuous liquid interface production	CLIP (Janusziewicz et al. 2016)
Digital light processing	DLP (Salonitis 2014)
Electrochemical additive manufacturing	ECAM (Kamraj et al. 2016)
Electron beam additive manufacturing (wire fed)	EBAM (Tarasov et al. 2019)
Electron beam melting	EBM (Korner 2016)
Fused deposition modeling	FDM (Masood 2014)
Fused pellet modeling	FPM (Wang et al. 2016)
High speed sintering	HSS (Brown et al. 2018)
Ink jet printing	IJP (Stringer and Derby 2009)
Laser engineered net shaping	LENS (Yan et al. 2018)
Localized microwave heating based AM	LMHAM (Jerby et al. 2015)
Micro droplet deposition manufacturing	MDDM (Chao et al. 2012; Zuo et al. 2016)
Microheater array powder sintering	MAPS (Holt et al. 2018)
Multi-jet fusion	MJF (Sillani et al. 2019)
Plasma arc additive manufacturing	PAD (Feng et al. 2018)
Photopolymer jetting	PJ (Lanceros-Méndez and Costa 2018)
Powder melt extrusion	PME (Boyle et al. 2019)
Rapid freeze prototyping	RFP (Bryant et al. 2003)
Selective heat sintering	SHS (Baumers et al. 2015)
Selective laser melting	SLM (Kumar 2014)
Selective laser sintering	SLS (Kumar 2003)
Stereolithography	SL (Salonitis 2014)
Thermoplastic 3D printing	T3DP (Scheithauer et al. 2017)
3D gel-printing	3DGP (Ren et al. 2016)
Two-photon polymerization	2PP (Nguyen and Narayan 2017; Wu et al. 2006)
Wire arc additive manufacturing	WAAM (Tabernero et al. 2018; Cunningham et al. 2018)

2.2.1 Materials

Processes can differ from each other on the basis of the materials they process. For example, fused deposition modeling (FDM) uses polymers or polymer based materials, and FDM differs from other processes because it works with polymers. The concept utilized in FDM is not unique to polymers, the concept can be used for metals as well; a metal wire instead of polymer filament can be taken instead, a high-temperature heater instead of a low-temperature heater will be used, some machine parts need to be changed and parameters need to be optimized. Thus, a process – FDM for metals, free from any polymer – cannot be an impossibility; this process needs to compete with other metal based processes technically and economically in order to be viable; it is perfectly impossible to predict perfect impossibility and perfect unviability of this process. But, predictability of FDM does not change the present fact that FDM is related to polymer. FDM is not free from polymer. It is not wrong to identify FDM with polymers, but it is certainly wrong to identify FDM with metals; expecting a polymer part from FDM is not expecting a wrong material part from FDM, but expecting a metallic part from FDM is nothing but expecting a wrong material part from FDM. Similarly, other AM processes can be distinguishable on the basis of materials such as photopolymer and only photopolymer is related to stereolithography (SL), metals can be identified with wire arc additive manufacturing (WAAM), metals can again be identified with additive friction stir deposition (AFSD) etc.

Thus, a single material can be used to identify a process, but a single material cannot be used to identify each and every process; a single material can be used to identify some processes while more than a single material is required to identify some other processes. For example, selective laser sintering (SLS) works with polymers, ceramics, metals and composites; this process cannot be identified with a single material (polymer or ceramic or metal), all materials are required. If all materials are required or if so many materials are required then a material, as an identifying probe, is no longer effectively identifying a process. Thus, a material has limited applicability to distinguish a process.

2.2.2 Agent for Joining Materials

Materials are joined by various means such as application of high-energy beam (laser, electron, plasma), other thermal source, low temperature, friction energy, binding agent, catalyst, kinetic energy, electrochemical energy etc. Some examples of the process related to these means or agents are given below:

1. **Laser**: it is used as a thermal source, for example SLS, SLM, LENS, CLF etc. It is used as a source of photons, for example SL, PJ etc. (Schmidt et al. 2017).
2. **Electron beam**: it is used as a thermal source, for example EBM, EBAM.
3. **Plasma beam**: it is used as a thermal source, for example PAD. It is generally obtained from Tungsten arc welding equipment.
4. **Other thermal source**: Micro heater, arc welding are used as other non-beam thermal sources, for example MAPS, WAAM etc.
5. **Friction energy**: it is used in AFSD etc.
6. **Kinetic energy**: it is used in CSAM.
7. **Binder**: it is used in IJP, BJ3DPetc.
8. **Catalyst**: it is used in 3DGP.

9. **Electrochemical energy**: it is used in ECAM, electrophoretic deposition for AM (Mora et al. 2018).
10. **Low temperature**: not very high temperature provided by a heat source is required for example FDM.
11. **Negative temperature**: it is used in RFP, cryogenic prototyping (Pham et al. 2008).

The above list is not exhaustive. Moreover, some processes use more than one means of joining, while the above list gives that one which is responsible for shaping. For example, ceramic laser fusion (CLF) uses both binder and laser beam, but it is the laser beam which is responsible for shaping, thus, CLF is included under laser beam. Another example is electron beam melting (EBM) which uses electron beam and high platform temperature, but it is the electron beam that is responsible for shaping, thus, EBM is included under electron beam.

2.2.3 Form of Feedstock

A feedstock is a material that is fed into a machine; it is a material that is going to be fed into the machine but has not yet been fed; when it will be fed into the machine, it will change, if machine is hot then it will become hot, if machine is cold then it will become cold, and if it changes its condition after being inside the machine, it is no longer the same material when it was waiting outside the machine. It is going to be fed into the machine and it is near to the machine but it is not far away from the machine. Because, when it will be far away from the machine, it can be corroded, it can be oxidized, it can absorb water and it can absorb other gases before it is brought near to the machine. Which distance is near and which one is far is material- and machine-dependent and is not certain, but one thing is certain that the feedstock – remained inside a machine (PBF, SL) when processing took place – does not give the same properties (Kumar and Czekanski 2018).

The following are the forms of feedstock used in AM: powder, wire, liquid, slurry, ion, gas; the form provides a distinctive mark on a process, for example, powder bed fusion (PBF) is distinguishable because it uses a particular form of material, that is powder (Snow et al. 2019), the importance of the form is more pronounced when it is found that other forms of solid such as wire, sheet, pellet, block etc. cannot be fitted in the concept of this process. Though, there are processes such as directed energy deposition (DED) or arc welding based AM which use more than one form of solid – powder and wire; but, using two forms does not trivialize the importance of a form; on the contrary, it emphasizes the uniqueness of a form – powder is used to furnish accuracy while wire is used to furnish high deposition rate, it demonstrates that the selection of a form is governed by the purpose of a process or the requirement of a part. Thus, there are two processes (DED-powder and DED-wire) emanating from one process (DED) because of a difference in the forms of the feedstock.

2.2.4 Conveyance of Feedstock

The method of transportation of same form of feedstock to the point of processing creates a difference between two processes. For example, powder is a form of feedstock for two different processes, that is PBF and DED, but in case of PBF, powder is brought by coating on a substrate or on a platform (Snow et al. 2019); while in DED, powder is brought on a substrate by blowing the powder via nozzle. Two processes (PBF and DED) are different because of a difference in conveyance of powder (a feedstock), but this difference is not only a difference in conveyance, this difference gives rise to a difference in timing of processing. In PBF, powders are conveyed; after the conveyance is over, these powders are processed either by jetting of binder or using a high-energy beam. In PBF, conveyance of powders results in the formation of a powder bed, powders lie still before processing happens. In PBF, timing of processing follows the conveyance of powders. In DED, powders are conveyed and are processed simultaneously by a high-energy beam; powders are blown, and at the same time the beam remains on. In DED, processing might not exactly coincide with the blowing of powders but, in principle, timing of processing does not follow the blowing of powders. In DED, processing does not wait for the blowing to be over before the processing starts. How powders are conveyed in two processes determines when the powders will be processed. In PBF, there is a cyclic sequence – powders are conveyed then powders are processed then powders are conveyed then powders are processed. In PBF, there is no processing when powders are conveyed or moved; and when they are not conveyed, only then they are processed or supposed to be processed. In DED, there is no such cyclic sequence, processing happens when powder is conveyed or blown or moved. In one process (PBF), processing happens when feedstock (powder) does not move, in another process (DED), processing happens when feedstock (powder) moves. Here, processing implies a transformation of powders into a structure, processing implies joining or an attempt of joining of powders. It needs to be checked whether the difference between two processes in terms of timing of processing related to the movement of feedstock is confined only to these two processes or is generic.

Stereolithography (SL) and photopolymer jetting (PJ) both use same form of feedstock, that is photopolymer, but these processes differ because photopolymer is conveyed differently. In SL, it is coated while in PJ it is jetted. In SL, photopolymer is cured after it is coated. In PJ, photopolymer is cured after it is jetted, but it needs to be cured immediately after it is jetted because delay in curing may lead to flowing or displacement or shape deformation of jetted drops. In SL, there is no such immediacy, there are no jetted drops that are about to leave their positions unless confined by curing, a coated layer can wait longer before getting cured. The difference between SL and PJ is not only in transportation of feedstock but also in timing of processing (curing) of feedstock after the transportation. In SL, processing happens when feedstock does not move; in PJ, processing happens when feedstock has just stopped moving. There is a clear difference between SL and PJ in timing of processing in relation to the movement of feedstock. This clear difference between SL and

PJ is not as much clear and accentuated as happened between PBF and DED. But, there is continuity in such differences, there is a trend in such differences, these differences may be generic.

Ceramic laser fusion (CLF) and thermoplastic 3D printing (T3DP) both use slurry as a feedstock; the difference between CLF and T3DP is the difference in conveyance of feedstock. In CLF, slurry is coated while in T3DP it is deposited. In CLF, coated slurry or a layer is processed or shaped by a laser beam. In T3DP, deposited slurry is processed by drying; here, drying is processing; this processing is not the only processing that is performed in this process, but this processing is sufficient enough to allow next round or next layer of deposited slurry to be deposited, this processing is expected to be fast lest the deposited slurry be deformed, this processing needs to be faster if the deposited slurry is of lower viscosity. In T3DP, there is urgency in drying. In CLF, there is no such urgency in processing; coated slurry can remain unprocessed for longer duration.

There is certainly a difference in CLF and T3DP for conveying feedstock, but, this difference does not end here, this difference leads to a difference in timing of processing after conveying. This difference in timing is not a new difference and this difference has been observed earlier, this difference has been observed in SL and PJ, this difference has also been observed in PBF and DED. There is a relation between CLF and T3DP in terms of similarity of feedstock material, in terms of conveyance of feedstock, in terms of the difference created. In these terms, there is also a relation between SL and PJ. In these terms, there is also a relation between PBF and DED. Thus, there is a relation each between CLF and T3DP, between SL and PJ, between PBF and DED. Thus, there are so many relations. But, these relations tend to obfuscate stronger relations existing among them. Observation of these relations tends to suggest that CLF, SL and PBF can be bracketed together because the common operation among them is that there is a substantial delay in processing after the conveyance of feedstock; on the basis of this common operation there exists a relation among them. While T3DP, PJ and DED can be bracketed together because there is no such substantial delay in processing after the conveyance of feedstock; on the basis of this commonality there exists a relation among these processes as well. Thus, there are two new relations. Observation suggests that there are no more relations. Thus, all six AM processes yet observed can be divided into two major categories, each category for each bracket or each new relation. The division of six AM processes into two major categories suggests a method to divide all AM processes into two major categories. This method has potential to classify all AM processes.

2.3 Classification by ASTM

ASTM (ASTM 2012) classified AM processes into seven categories, as displayed in Table 2.2.

Table 2.2 Classification of AM processes as per ASTM

Category	Process
1	Binder jetting. Example: BJ3DP
2	Directed energy deposition. Examples: LENS, EBAM
3	Material extrusion. Example: FDM, FPM, PME
4	Material jetting. Examples: IJP, PJ
5	Powder bed fusion. Examples: SLS, SLM, EBM
6	Sheet lamination. Examples: UC (White 2003)
7	Vat photopolymerization. Examples: SL, DLP

2.3.1 Complication with ASTM Classification

There are the following complications with the classification and categories.

2.3.1.1 Exclusion of Many Processes

There are many AM processes that could not find a place in any category. For example, CSAM (cold spray does not come under material jetting), ECAM, CLF (slurry layering does not come under binder jetting or powder bed fusion), WAAM (arc of the arc welding does not come under focused thermal energy and will not come under the second category, i.e. directed energy deposition).

2.3.1.2 Inability to Suggest

The classification is not able to suggest what will be the eighth category.

2.3.1.3 Sheet Lamination

This category is not an AM as per the definition provided by ASTM. It is a contradiction that sheet lamination that is not an AM is part of a classification meant solely for AM processes (details in Chap. 11).

2.3.1.4 Vat Photopolymerization

The name of this category is misnomer. There is no need for a vat for vat photopolymerization to occur. Even a plate without a confining or boundary wall is sufficient for these processes to occur. For example, if vat photopolymerization of the type of inverse stereolithography (Hafkamp et al. 2017; Chi et al. 2013) is selected for processing a high-viscous photopolymer, then a vat is not required; instead, a plate can serve the purpose. The plate can be coated with a layer of high-viscous photopolymer, a confining wall is not required to contain the photopolymer on the plate, since the high viscosity of the photopolymer will not let it spread and change the layer thickness before it is cured by a fast beam. Thus, a plate is sufficient for vat photopolymerization to be accomplished; in this case, plate photopolymerization will be more appropriate term than vat photopolymerization. Thus, the name of the category does not adequately represent the processes involved (details in Chap. 8).

2.4 Attempt to Classify on the Basis of Materials

If processes will be classified on the basis of materials such as metals, polymers, ceramics and composites, then it will give guarantee that all processes get classified because there is not a single process which does not deal with any material. For any material, there are a number of AM processes to process that material. For example, for processing metals, the processes available are LENS, DED-wire, SLS, SLM, EBM etc.; thus, there are a number of processes available which cannot be separated from each other because they all process metals (the same material). If a metal is further divided into various groups such as iron based alloy, titanium, nickel etc., there are again a number of processes available which process a particular metal or metallic alloy. It does not imply that if the same type of metallic alloy will be processed by these AM processes, they all will not give different metallic properties. Even if these processes give different metallic properties, it will not lead them to have different places in the classification because the classification is on the basis of materials and not on the basis of material properties. If material properties are not excluded because material plus material properties have better prospect to become a basis for classification, and this new basis is giving different places for different AM processes in the classification, then a particular AM process will be having not only a single different place but also many different places, because a particular AM process will process many materials which will result in many specific properties. It will lead far away from an ideal goal – a single place for one process in the classification.

There is not a single metallic alloy available which is processed by a dedicated single process, and that dedicated single process is not supposed to process any other metals other than this single metallic alloy. In other words, there are not specific materials available each of which corresponds to a single process. Consequently, if classification is done on the basis of materials, then there are no materials available to distinguish between two AM processes; materials are not capable to distinguish between two AM processes because processes available are not amenable to be distinguished by materials.

2.5 Attempt to Classify on the Basis of Agents for Joining Materials

If the classification is done on the basis of agents for joining materials such as laser beam, electron beam, binder, kinetic energy, low temperature, microwave energy etc. then there are some AM processes which will be well placed in the classification. One example is CSEM because CSEM is a single example of its corresponding agent, that is kinetic energy. In the case of this agent, there is no need to sub-classify as there is not more than one process to be accommodated. Consequently, single agent-single process type case will suit well in this classification.

However, if any agent has more than one processes, for example laser has many processes such as SLS, LENS, SLM, SL, PJ etc. (Schmidt et al. 2017), then further categories of laser need to be searched to find individual place for these processes. If laser is categorized as high-, medium- and low-power laser, then processes will be categorized as high (LENS, SLM), medium (SLS) and low (SL, PJ). This categorization may not be strict because SLS and SLM have overlapping laser power, but it needs to be checked even if the processes are categorized in this way where this classification will lead. Further, this categorization of laser could not help LENS and SLM to be separated. If the laser is further sub-categorized as pulse and continuous mode, or Nd: YAG, CO_2, fibre type, then still both processes (LENS, SLM) could not be separated. There is no attribute of laser available which is capable to separate LENS from SLM (Schmidt et al. 2017). Thus, this classification type has limitations; this classification is suitable if there is one or few processes belonging to an agent; this is not suitable to accommodate increasing number of AM processes. In other words, if AM succeeds to have a number of processes for each type, this classification fails.

2.6 Attempt to Classify on the Basis of Form of Feedstock

If the classification is done on the basis of form of feedstocks, then there will be following sub-headings in the classification: powder, wire, gas, liquid, gel, slurry etc. In order to check how this classification will fare – powder as an example of the form of feedstock is taken. Powder can be further divided into following categories: metal powder, polymer powder, ceramic powder. Under the sub-heading of metal powder, following AM processes will come: SLM, LENS, CSEM, MDDM, EBM, AFSD, Arc welding based processes etc. There are so many processes under the sub-heading named metal powder. There are no more attributes of powder (such as small, big, high surface roughness, low surface roughness, wide powder size distribution, narrow powder size distribution, well-flowing, not-well-flowing, spherical shape, non-spherical shape, fully dense, porous etc.) which will help these processes to be further categorized. This brings an end to this classification; there will not be any more branching; there is no way to prevent these processes to be clubbed together. If so many processes will be categorized as just one type of process, then this classification does not help in selecting one metal powder type AM process from other metal powder type AM process. Metal powder is a common feature among these processes but is not the dominant feature; it is not so dominant that it will guide majority of other features of the processes to be common. Clubbing together these processes might be making sense if not clubbing them together might not be making sense, for example, clubbing together SLS and SLM on the basis of powder makes more sense than not clubbing together because many common features they do have.

The form as a basis of classification does not help do classification, it also does not lead to inherent characteristics of the processes which can become a basis for the classification.

2.7 Attempt to Classify on the Basis of Conveyance of Feedstock

There are many ways a feedstock is conveyed, such as coating (SL, SLS, EBM, SLM, BJ3DP, CLF etc.), blowing (LENS, CSEM, powder based arc welding process etc.), powder feeding (AFSD), material jetting (IJP, PJ, MDDM etc.), air jetting (AJ), extruding (FDM etc.), wire feeding (DED-wire, WAAM etc.), no feeding (2PP, CLIP etc.) etc. From the perspective of conveyance of feedstock, AM processes can be classified into two categories: no feeding category and feeding category.

2.7.1 No Feeding Category

The classification based on the conveyance of feedstocks needs to take into account those AM processes where conveyance of feedstocks does not occur. It is because conveyance is not required; materials are not required to be fed because materials are already there. For example, in case of 2PP and CLIP photopolymer in a container is already present to carry out the process; these processes are not waiting for photopolymer to be either coated or jetted for the next task to perform; feeding (the photopolymer) is not a component of these processes. It does not imply that the container will not be fed with the photopolymer in the beginning, at the end, or the container will not be periodically replenished; but this type of feeding of the container or filling up the container is not the same as the feeding which happens to be a periodic or a recurring step of a process, the feeding which is used to bring a predetermined amount of materials, at a certain time, at a certain point in a certain fashion without which the process refuses to proceed.

The common thing among all these processes is that there is no movement of feeding materials, materials remain still or motionless; these processes can be combinedly called motionless material process. Here, 'motionless' implies that there is no motion of materials due to feeding, it does not imply that there will be no motion due to solidification or phase transformation.

2.7.2 *Feeding Category*

There are few processes which do not require feeding, majority of AM processes barring few require feeding. Those processes which require feeding or which require some form of conveyance of feedstocks can be divided into two major categories as follows:

1. Coating
2. Blowing, powder feeding, wire feeding, material jetting, air jetting, extruding

The first category is different from the second category. In the first category, there is a delay between feeding and its transformation; in the second category, there is no such delay between feeding and its transformation. In the first category, transformation starts after the coating of photopolymer or powder or slurry is done. In the second category, transformation does not wait for the blowing of powder to be over, the transformation concurs with the blowing of powder; the transformation concurs with the feeding of wire. In the second category, the transformation does not wait so long when material is jetted or extruded; in the first category, transformation waited so long when the material was coated; waiting in the first category and lack of such waiting in the second category do not connote to the inertia of the respective machines; these waitings or no waitings are the necessary steps, requirements or the inherent characteristics of the processes.

2.7.2.1 Feeding Category of the First Type

The first category is related to the coating of the feedstock. In this category, materials are placed on a substrate or on a platform. In case of SLS, SLM and EBM, materials in the form of powders are placed, placing the powders is also akin to creation of a powder bed; these processes are thus also named as powder bed fusion. In BJ3DP, a powder bed is also formed while in CLF a slurry bed is formed. In SL, photopolymer is coated or photopolymer is placed which is akin to creating a photopolymer bed.

In all AM processes related to coating of feedstock, a material bed is formed; this material bed is either from powder or slurry or photopolymer. AM processes belonging to the first type can thus be combinedly called material bed process.

Material bed process can thus be classified as solid bed and liquid bed as shown in Fig. 2.1. Solid bed, which is usually powder bed, can be further classified as powder bed fusion and powder bed non-fusion as shown in Fig. 2.2; fusion implies melting, powder bed fusion can be classified as complete fusion based and partial fusion based depending upon the complete melting and partial melting of powders respectively. Melting is induced either by a laser beam or by an electron beam, thus complete fusion can be classified as laser beam based and electron beam based. SLM belongs to complete fusion type, EBM belongs to complete fusion induced by electron beam type, SLS and HSS belong to partial fusion type. BJ3DP belongs to non-fusion type.

Powder bed fusion can also be classified as beam based PBF and non-beam (or heater/lamp) based PBF; in beam based PBF, SLS, SLM and EBM will come. Beam based PBF can be further divided into two types: laser beam PBF or laser PBF (comprising of SLS and SLM) and electron beam PBF (comprising of EBM). Laser PBF is given in Chap. 3 while electron beam PBF is given in Chap. 4. In non-beam based PBF, high speed sintering (HSS) and multi-jet fusion (MJF) will come; it shows there are many ways a sub-classification can be accomplished (details in Chap. 3).

Liquid bed can be classified as photopolymer bed and slurry bed. Photopolymer can be further classified as scan based and projection based (DLP) as shown in Fig. 2.3. Complete classification of material bed process combining Figs. 2.1, 2.2 and 2.3 is given in Fig. 2.4.

2.7.2.2 Feeding Category of the Second Type

The second category is related to the following type of feeding: blowing, wire feeding, jetting, extruding. In all feeding type, material moves from one point to another to make a structure; in blowing, powder is moved from a nozzle to a substrate to make a structure; in wire feeding, wire moves from feeder to the substrate to make a structure; in jetting, liquid moves from printhead to the substrate to make a structure; in extruding, polymer moves from a nozzle to the substrate to make a structure. This is different from the previous category where in coating, powder or slurry or photopolymer moves from one point to another to make a bed; they do not make any structure, instead they always make same type of bed. While in this category, material is moved to make a design, material is deposited to create a design; this is

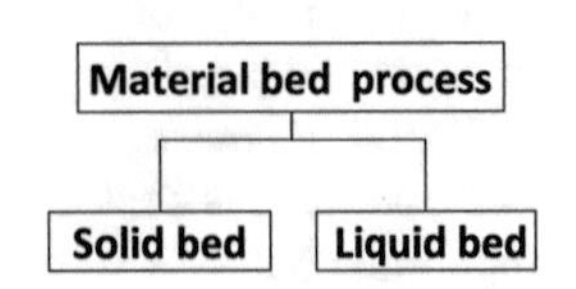

Fig. 2.1 Classification of material bed process into two major types

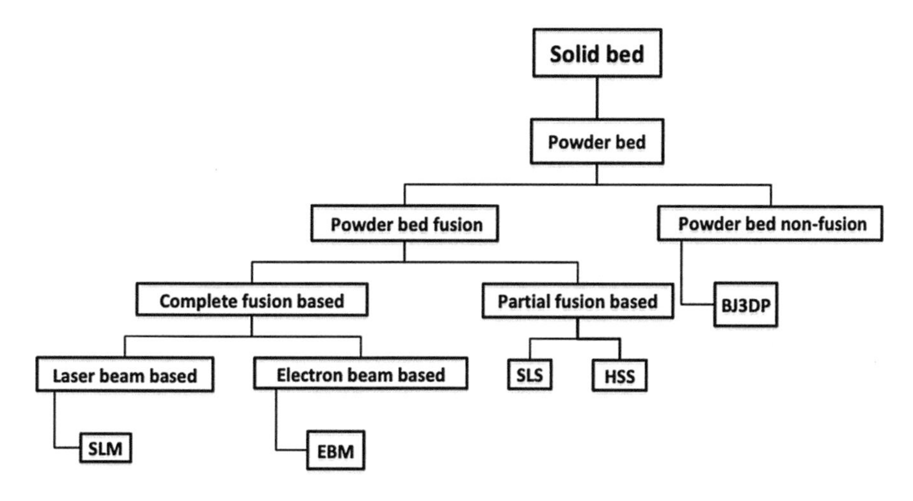

Fig. 2.2 Classification of solid bed process

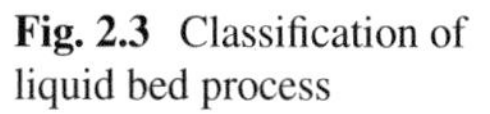

Fig. 2.3 Classification of liquid bed process

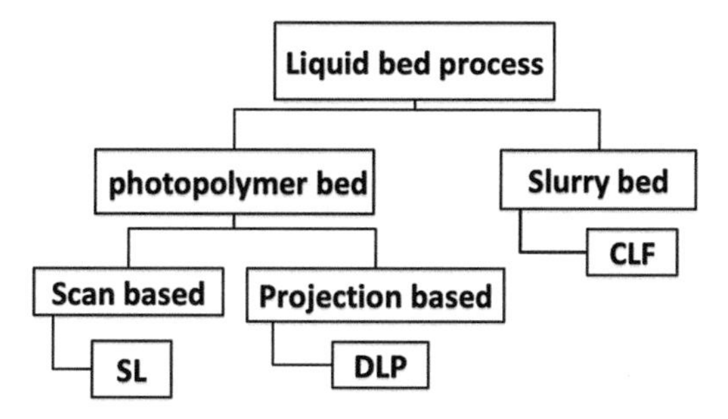

common among all processes in this category, there is no exception. Therefore, all processes can be combinedly called material deposition process.

The second category or material deposition processes can be classified into following categories: solid deposition, liquid deposition, air deposition and ion deposition as shown in Fig. 2.5. Solid deposition can be further classified as powder deposition, wire deposition (Fredriksson 2019), filament deposition and rod deposition as shown in Fig. 2.7. Powder or wire can be deposited by various sources of energy such as laser, electron beam, plasma, arc welding, friction, cold spray etc. (Dass and Moridi 2019). Powder deposition can thus be classified as laser based (LENS), plasma based, friction based (AFSD), cold spray based (CSAM) and arc welding based. Wire deposition can be classified as laser based, electron beam based (EBAM), plasma beam based, arc welding based (WAAM). Liquid deposition can be classified as polymer deposition, ink deposition (IJP), photopolymer deposition (PJ), metal deposition, water deposition (RFP), slurry deposition as shown in Fig. 2.7; while slurry can be expanded as photopolymer based slurry, polymer based slurry (T3DP) and gel based slurry (3DGP) as shown in Fig. 2.8. Air

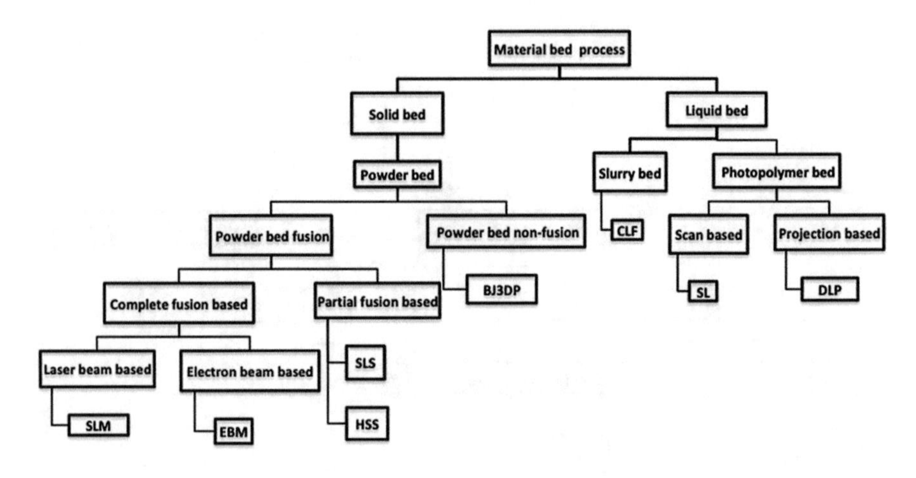

Fig. 2.4 Classification of material bed process

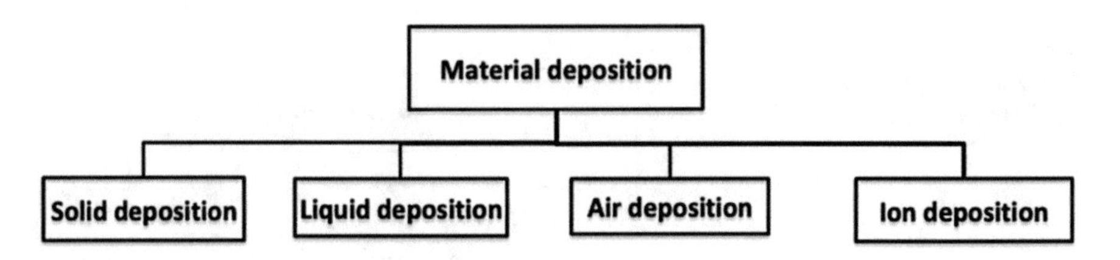

Fig. 2.5 Classification of material deposition process

deposition is aerosol jetting (AJ) while ion deposition is electrochemical additive manufacturing (ECAM).

2.8 Difference Between Solid Deposition Process and Liquid Deposition Process

A solid deposition process (SDP) uses solid feedstock while a liquid deposition process (LDP) uses liquid feedstock. For example, LENS as an SDP uses powder as solid feedstock while IJP as an LDP uses ink as liquid feedstock. Since most of the SDP makes products by changing from solid to liquid to solid, SDP requires parameter optimization both for control of flow of solid feedstock and control of liquid that forms. Since there is no solid in LDP, only control of liquid is required in LDP.

What if LDP uses solid feedstock as a source for liquid feedstock. For example, in metal jetting (Simonelli et al. 2019), it is not convenient for LDP system to store liquid

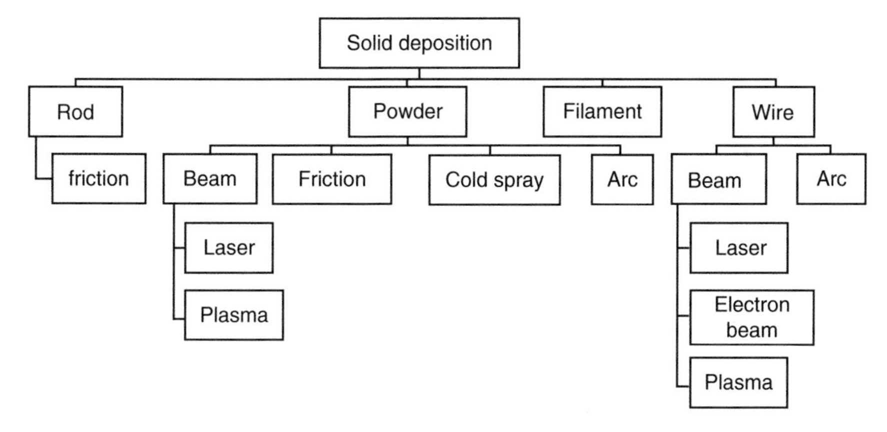

Fig. 2.6 Classification of solid deposition process

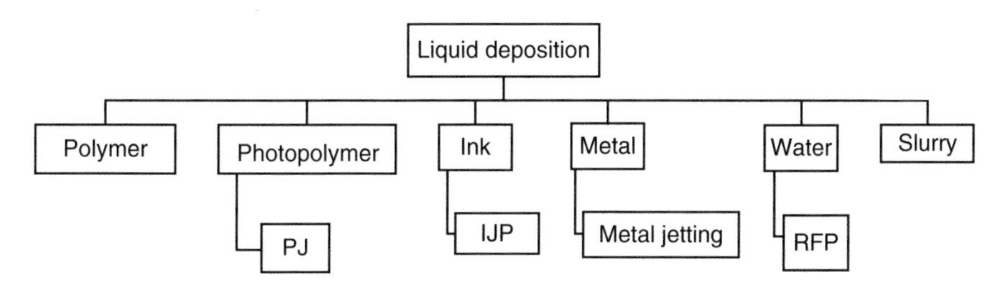

Fig. 2.7 Classification of liquid deposition process

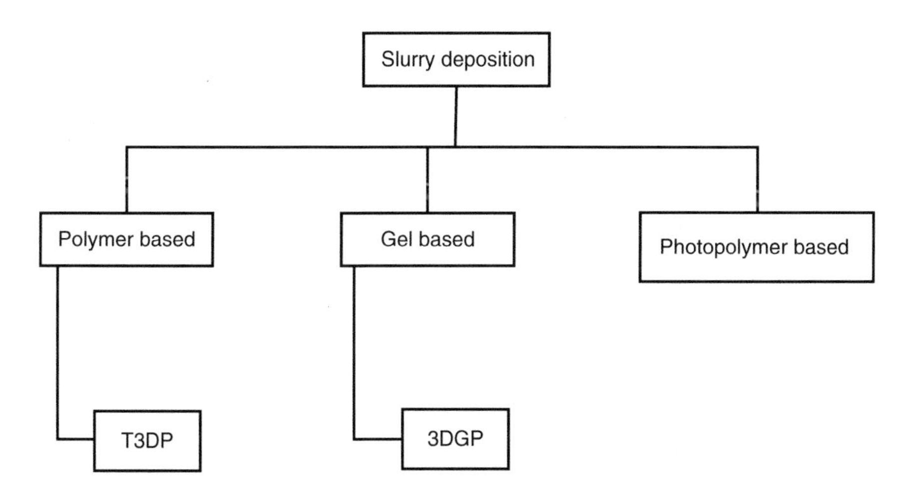

Fig. 2.8 Classification of slurry deposition process

feedstock or molten metal because then the system needs to be maintained at high temperature even if it is not working. In this example, therefore, it is more convenient to use solid feedstock such as metal powders or rods or shots and melt them to get liquid feedstock when required. This example will still come under LDP, though it uses solid feedstock because the process is controlled by controlling the amount and frequency of liquid jet; solid feedstock has no direct connection with the substrate and controlling the amount of solid feedstock will not change the liquid deposition or the amount and frequency of liquid jet. It does not mean that solid feedstock has no impact on LDP – the role of solid feedstock is only to ensure that there is sufficient amount of liquid feedstock available which will be further controlled but the role of solid feedstock is not to provide a means to directly control the amount and frequency of liquid jet. In SDP, the role of solid feedstock is to provide a means to directly control the size and shape of deposited material; for example, in wire feeding EBM, by controlling solid feedstock rate (wire feed rate), the size and shape of molten pool can be controlled; for example, in FDM, the role of filament (solid feedstock) is to provide a means to directly control the size and shape of extruded material. This is how an LDP with solid feedstock differs from an SDP with solid feedstock.

2.9 Classification of AM

On the basis of feeding categories, AM processes can be classified into following three types (as shown in Fig. 2.8):

(i) Material bed process (from feeding category of the first type)
(ii) Material deposition process (from feeding category of the second type)
(iii) Motionless material process (from no feeding category)

Further classification of the first type, that is material bed process is given in Fig. 2.4 while further classification of the second type, that is material deposition process is given in Figs. 2.5, 2.6, 2.7 and 2.8.

On the basis of layerwise fabrication, AM can be classified into two types: additive layer manufacturing (ALM) and additive non-layer manufacturing (ANLM) (given in Chap. 1), while Fig. 2.9 states that AM can be classified into three types. It brings a vital question – which classification needs to be adopted.

On the basis of layerwise fabrication, AM can be seen as it is going through making layer or it is not going through making layers. But, it does not tell how layers are formed, it also does not tell if layers are not formed then how non-layers are formed – it is oblivious of the fact that there are feedstocks involved in the formation of layers or non-layers.

As per Fig. 2.9, AM can be seen in relation to feedstocks, Fig. 2.9 tells how materials are moving, how materials are getting deposited or how materials are not getting deposited but Fig. 2.9 is silent on the consequence of these movements or no movements in terms of formation of layers. The consequence might be the formation of layers or the formation of non-layers, but Fig. 2.9 is not giving those information. It could be better if Fig. 2.9 might be giving more information because more

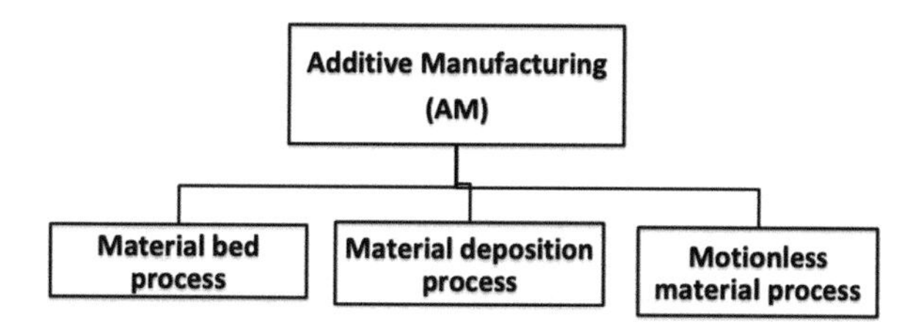

Fig. 2.9 Classification of AM on the basis of feeding

information will certainly provide more clarification and clear classification. Combining Fig. 2.9 with classification based on layerwise fabrication will give more information or required missing information. In the absence of this combination, there will be incomplete information.

Figure 2.10 is a result of this combination, in which each type of Fig. 2.9 is divided into two types, such as material bed process is of two types ALM and ANLM, material deposition process is of two types ALM and ANLM, motionless material process is of two types ALM and ANLM.

Material bed formation itself implies that a layer in the form of bed is formed and is obviously of ALM type. All material bed process types shown in Fig. 2.4 are of ALM types. Since all material bed processes are presently ALM types, it gives a wrong impression that a material bed process of ANLM type is an impossibility. Material bed process implies that materials are deposited in the form of a bed or in the form of a layer, it does not imply that materials should also be consolidated in the form of a layer. Since, without exception, in material bed process, materials are going to be deposited in the form of a layer, and then are going to be consolidated in the form of a layer; it has become a de facto rule. If materials will be deposited in the form of a layer but will not be consolidated in the form of a layer, then this process will be material bed process of ANLM type. For example, if twenty layers of materials are deposited but this heap of materials is consolidated sidewise (not from the top) to make some structure, afterwards ten layers are deposited which are consolidated again from another side to make another geometry; this will be an example of layerwise deposition and non-layerwise consolidation.

Material deposition process shown in Figs. 2.5, 2.6, 2.7 and 2.8 are of ALM type while layerless FDM (Kanada 2015) (given in Chap. 10) will come under material deposition process of ANLM type. Most of the ANLM such as 2PP, CLIP come under motionless material process. There is no material bed process of ANLM type and there is no motionless material type of ALM type, therefore these two categories can be removed from the classification (Fig. 2.10). The final classification thus derived is shown in Fig. 2.11. This classification accommodates all existing AM processes; and this classification consists of such broad varieties that yet-to-be-invented processes can also be accommodated. If yet-to-be-invented processes can be accommodated, then it brings a question where they are going to be accommodated. If it is known where they are going to be accommodated, then it is providing information what is going to be invented. Thus, there are possibilities to know future processes from the classification. Future processes based on the classification are given in Chap. 12.

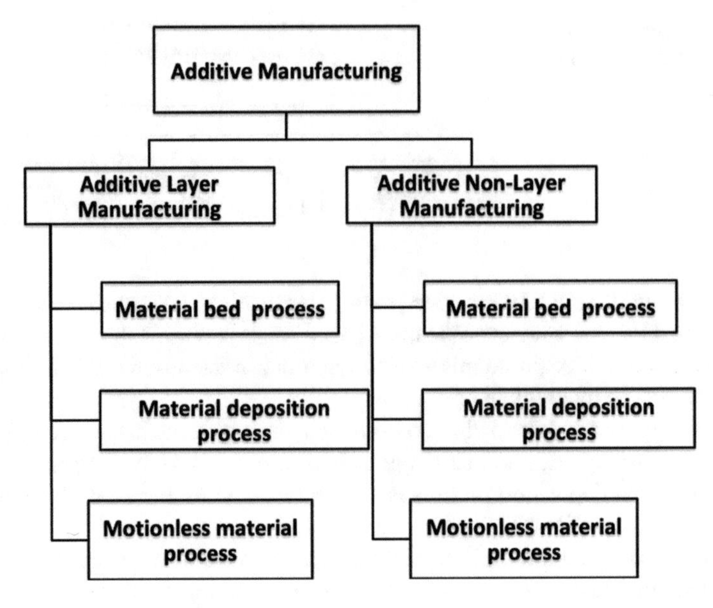

Fig. 2.10 Combined classification of AM

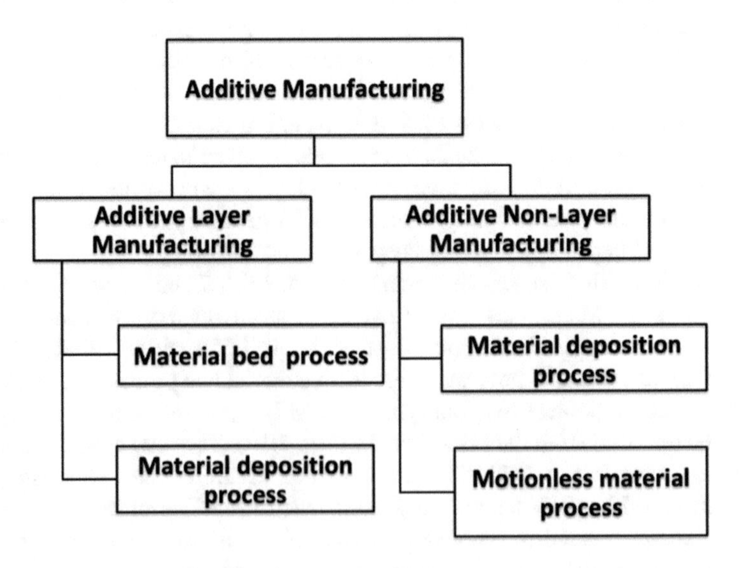

Fig. 2.11 Final classification of AM

References

ASTM F2792-12a (2012) Standard terminology for additive manufacturing technologies (withdrawn 2015). ASTM International, West Conschohocken

Baumers M, Tuck C, Hague R (2015) Selective heat sintering versus laser sintering: comparison of deposition rate, process energy consumption and cost performance. In: SFF proceedings, pp 109–121

Boyle BM, Xiong PT, Mensch TE et al (2019) 3D printing using powder melt extrusion. Addit Manuf 29:100811

Brown R, Morgan C T, Majweski C E (2018) Not just nylon—improving the range of materials for high speed sintering. In: SFF proceedings, pp 1487–1498

Bryant FD, Sui G, Leu MC (2003) A study on effects of process parameters in rapid freeze prototyping. Rapid Prototyp J 9(1):19–23

Chao Y, Qi L, Xiao Y et al (2012) Manufacturing of micro thin-walled metal parts by microdroplet deposition. J Mater Process Technol 212(2):484–491

Chi Z, Yong C, Zhigang Y, Behrokh K (2013) Digital material fabrication using mask-image-projection-based stereolithography. Rapid Prototyp J 19(3):153–165

Cunningham CR, Flynn JM, Shokrani A et al (2018) Invited review article: strategies and processes for high quality wire arc additive manufacturing. Addit Manuf 22:672–686

Dass A, Moridi A (2019) State of the art in directed energy deposition: from additive manufacturing to materials design. Coatings 9(418):1–26

Enneti RK, Prough KC, Wolfe TA et al (2018) Sintering of WC-12%Co processed by binder jet 3D printing (BJ3DP) technology. Int J Refract Met Hard Mater 71:28–35

Feng Y, Zhan B, He J, Wang K (2018) The double-wire feed and plasma arc additive manufacturing process for deposition in Cr-Ni stainless steel. J Mater Process Technol 259:206–215

Fredriksson C (2019) Sustainability of metal powder additive manufacturing. Procedia Manuf 33:139–144

Goh GL, Agarwala S, Tan YJ, Yeong WY (2018) A low cost and flexible carbon nanotube pH sensor fabricated using aerosol jet technology for live cell applications. Sensors Actuators B Chem 260:227–235

Hafkamp T, Baars G V, Jager B D, Etman P (2017) A trade-off analysis of recoating methods for vat photopolymerization of ceramics. In: SFF proceedings, vol 28, pp 687–711

Holt N, Horn AV, Montazeri M, Zhou W (2018) Microheater array powder sintering: a novel additive manufacturing process. J Manuf Process 31:536–551

Janusziewicz R, Tumbleston JR, Quintanilla AL et al (2016) Layerless fabrication with continuous liquid interface production. PNAS 11(42):11703–11708

Jerby E, Meir Y, Salzberg A et al (2015) Incremental metal-powder solidification by localized microwave-heating and its potential for additive manufacturing. Addit Manuf 6:53–66

Johannes S J, Keicher D M, Lavin J M et al (2018) Multimaterial aerosol jet printing of passive circuit elements. In: SFF symposium proceedings, pp 473–478

Kamraj A, Lewis S, Sundaram M (2016) Numerical study of localized electrochemical deposition for micro electrochemical additive manufacturing. Procedia CIRP 42:788–792

Kanada Y (2015) Support-less horizontal filament stacking by layer-less FDM. In: SFF proceedings, pp 56–70

Kernan BD, Sachs EM, Oliveira MA, Cima MJ (2007) Three dimensional printing of tungsten carbide-10 wt % cobalt using a cobalt oxide precursor. Int J Refract Met Hard Mater 25:82–94

Korner C (2016) Additive manufacturing of metallic components by selective electron beam melting- a review. Int Mater Rev 61(5):361–377

Kumar S (2003) Selective laser sintering-a qualitative and objective approach. JOM 55(10):43–47

Kumar S (2014) Selective laser sintering/melting. Compr Mater Process 10:93–134. Elsevier Ltd

Kumar S, Czekanski A (2018) Roadmap to sustainable plastic additive manufacturing. Mater Today Commun 15:109–113

Lanceros-Méndez S, Costa CM (2018) Printed batteries: materials, technologies and applications. Wiley, Hoboken
Masood SH (2014) Advances in fused deposition modeling. Compr Mater Process 10:69–91
Mora J, Dudoff JK, Moran BD et al (2018) Projection based light-directed electrophoretic deposition for additive manufacturing. Addit Manuf 22:330–333
Nguyen AK, Narayan RJ (2017) Two-photon polymerization for biological applications. Mater Today 20(6):314–322
Pham CB, Leong KF, Lim TC, Chian KS (2008) Rapid freeze prototyping technique in bio-plotters for tissue scaffold fabrication. Rapid Prototyp J 14(4):246–253
Ren X, Shao H, Lin T, Zheng H (2016) 3D gel-printing- an additive manufacturing method for producing complex shaped parts. Mater Des 101:80–87
Roschli A, Gaul KT, Boulger AM et al (2019) Designing for big area additive manufacturing. Addit Manuf 25:275–285
Salonitis K (2014) Stereolithography. Compr Mater Process 10:19–67. Elsevier
Scheithauer U, Potschke J, Weingarten S et al (2017) Droplet-based additive manufacturing of hard metal components by thermoplastic 3D printing (T3DP). J Ceram Sci Technol 8(1):155–160
Schmidt M, Merklein M, Bourell D et al (2017) Laser based additive manufacturing in industry and academia. CIRP Ann 66(2):561–583
Sillani F, Kleijnen RG, Vetterli M et al (2019) Selective laser sintering and multi jet fusion: process-induced modification of the raw materials and analyses of parts performance. Addit Manuf 27:32–41
Simonelli M, Aboulkhair N, Rasa M et al (2019) Towards digital metal additive manufacturing via high-temperature drop-on-demand jetting. Addit Manuf 30:100930
Snow Z, Martukanitz R, Joshi S (2019) On the development of powder spreadability metrics and feedstock requirements for powder bed fusion additive manufacturing. Addit Manuf 28:78–86
Stringer J, Derby B (2009) Limits to feature size and resolution in ink-jet printing. J Eur Ceram Soc 29:913–918
Tabernero I, Paskual A, Alvarez P, Suarez A (2018) Study on arc welding processes for high deposition rate additive manufacturing. Procedia CIRP 68:358–362
Tang HH (2002) Direct laser fusing to form ceramic parts. Rapid Prototyp J 8(5):284–289
Tarasov SY, Filippov AV, Shamarin NN et al (2019) Microstructural evolution and chemical corrosion of electron beam wire-feed additively manufactured AISI 304 stainless steel. J Alloys Compd 803:364–370
Wang Z, Liu R, Sparks T, Liou F (2016) Large scale deposition system by an industrial robot (I): design of fused pellet modeling system and extrusion process analysis. 3D. Print Addit Manuf 3(1):39–47
White D (2003) Ultrasonic object consolidation. U.S. Patent No. 6,519,500. Washington, DC
Wu S, Serbin J, Gu M (2006) Two-photon polymerization for three-dimensional micro-fabrication. J Photochem Photobiol A Chem 181:1–11
Yan Z, Liu W, Tang Z et al (2018) Review on thermal analysis in laser-based additive manufacturing. Opt Laser Technol 106:427–441
Yin S, Cavaliere P, Aldwell B et al (2018) Cold spray additive manufacturing and repair: fundamentals and applications. Addit Manuf 21:628–650
Yu HZ, Jones ME, Brady GW et al (2018) Non-beam-based metal additive manufacturing enabled by additive friction stir deposition. Scr Mater 153:122–130
Zuo H, Li H, Qi L, Zhong S (2016) Influence of interfacial bonding between metal droplets on tensile properties of 7075 Aluminum billets by additive manufacturing technique. J Mater Sci Technol 32(5):485–488

Chapter 3
Laser Powder Bed Fusion

Abstract Both selective laser sintering (SLS) and selective laser melting (SLM) are laser powder bed fusion. This chapter provides working of SLS and SLM and the role of their various parameters. Fabrication speeds due to these processes are low which is one of the drawbacks of additive manufacturing; this chapter provides various methods to increase the speed. It is reasoned why there are only two binding mechanisms (liquid phase sintering and full melting) instead of four (liquid phase sintering, full melting, solid state sintering and chemical-induced binding) in laser powder bed fusion. Application of the process in the repair of a 3D part, though not commonly practiced, is given.

Keywords Laser · Sintering · Melting · Binding mechanism · Classification · Fabrication rate

3.1 Powder Bed Process

Powder bed process (PBP) is a generic name for additive manufacturing processes in which a powder bed is created and selectively joined to make a part. Powder bed means a thin layer of powders spread on a platform (or a substrate) or somewhere. This thin layer corresponds to a slice of a 3D CAD model of a would-be part. PBP implies thus creation of a powder bed and its consolidation (joining) thereafter. Variation in powder beds or variation in types of joining brings variation in PBP. For example, if a powder bed is compacted by applying pressure then it is no longer similar to a powder bed created by just spreading powders using a roller or a scraper (or a recoater); the compaction will thus create another PBP. In other examples of variation, if powders are joined by two types: (1) by completely melting them using a laser beam or (2) by interlocking them using binder; then these two types of joining are not same and will give rise to two types of PBP.

S. Kumar, *Additive Manufacturing Processes*,
https://doi.org/10.1007/978-3-030-45089-2_3

3.1.1 Classification

PBP is found to be of two major types: (1) powder bed fusion (PBF) and (2) powder bed non-fusion (PBNF). As per Oxford dictionary, fusion is the process of causing a material or object to melt with intense heat so as to join with another. In PBF, powders are partially or fully melted in order to join them, for example selective laser sintering (SLS) (Kumar 2010), selective laser melting (SLM), electron beam melting (EBM) (Korner 2016), selective heat sintering (SHS) (Baumers et al. 2015), micro heater array powder sintering (MAPS) (Holt et al. 2018), high speed sintering (HSS) (Brown et al. 2018) and localized microwave heating based additive manufacturing (LMHAM) (Jerby et al. 2015). In PBNF, powders are not melted, binders are used to join powders, for example binder jet 3D printing (BJ3DP) (Enneti et al. 2018), selective inhibition sintering (SIS) (Khoshnevis et al. 2014) etc.

PBF can be further divided into two categories: (1) beam based processes which require high-energy beams, for example SLS, SLM, EBM, and (2) non-beam based processes which do not require a high-energy beam but can work with other thermal sources such as heaters, lamps and microwave, for example SHS, HSS, MAPS and LMHAM. Beam based powder bed fusion can be further divided into two categories: (1) laser powder bed fusion comprising of SLS and SLM and (2) electron beam powder bed fusion comprising of EBM. This chapter deals with only laser powder bed fusion (LPBF), Chap. 4 deals with electron beam powder bed fusion (EPBF) while other powder bed processes are dealt with in Chap. 5.

Classification of PBP given in Fig. 3.1.

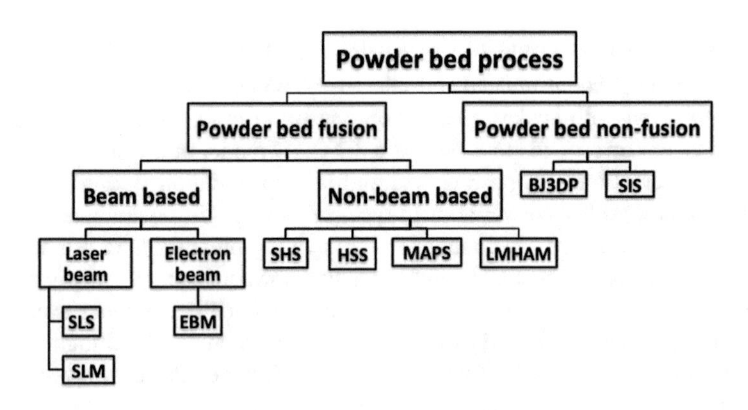

Fig. 3.1 Classification of powder bed process

3.1.2 Description of Classification

In SLS, if a polymer-coated high melting point material as a powder is used, then it is the polymer which is melted which helps join high melting point materials without fusing them; therefore, it is not a case of joining high-temperature materials by fusing or melting them but because polymer is melted or part of powder is melted or part of powder bed is melted, this case of SLS still comes under PBF.

In HSS, before powder bed is scanned by a thermal lamp, a radiation absorbing ink is deposited using an ink jet print head (Thomas et al. 2006); use of a print head similar to that used in BJ3DP gives an impression that HSS comes under PBNF, but the use of ink in HSS is not to bind powders but to facilitate subsequent melting and fusion of powder bed, HSS thus comes under PBF.

In BJ3DP, binder is deposited on the powder bed using an ink jet print head; the job of the binder is to hold powders together, but if the binder is not capable to do that, then a thermal lamp is used to improve the efficiency of binders; the lamp is not meant to melt powders (Liravi and Vlasea 2018). SIS is same as BJ3DP in terms of requirement of a jet print head, where inhibiter replaces the binder as a depositing material; the role of inhibiter is to act as a negative binder. If one process (SIS) uses negative binder and other (BJ3DP) uses positive binder, then both processes can also be named as binder jetting processes, but this new name is not better name than the previous name (powder bed non-fusion) for the purpose of classification. Because a new name will not prevent another process (slurry bed) to be classified together with SIS and BJ3DP, though slurry bed process does not use powder (given in Chap. 8).

3.1.3 Role of Heat

Almost all PBPs utilize heat as a basic necessity to realize the process. In BJ3DP, while binder jetting is used to shape a part, heat is utilized in post-processing to sinter the part. Similarly, in SIS, heat is used as a necessity in the form of post-processing sintering; in absence of post-processing, part will never form. In SLS, SLM and EBM, heat is used both for fabricating and post-processing. In SLM and EBM, controlled heat in the form of a point source is only a necessity to fabricate a part; post-processing is used to improve properties but is not always a necessity. In SLS, when point heat source is used for shaping a part, post-processing heat treatment is always a necessity to make the part usable; in this case unlike SLM and EBM, employing heat as post-processing is always a necessity.

3.2 Laser Powder Bed Fusion

Both selective laser sintering and selective laser melting are laser powder bed fusion (LPBF). The process traces its origin back to 1986 when Carl Deckard, a master's student of University of Texas, filed a patent on this process; the patent was granted in 1989. The process was later commercialized by DTM corporation (now 3D System) and since then the process has seen exponential growth in all aspects: related patents filed, types of materials processed, papers presented and published, new applications found, machines sold, new industries which adopted it for research and production etc.

The term 'laser' in selective laser sintering and selective laser melting implies that a laser is used for processing; the term 'sintering' implies that powders are involved in the process. It infers that powder processing is done by the laser to make parts. It could only be possible if the laser is used as a heat source. 'Selective' implies that all powders are not processed by the laser simultaneously, or in other words powders are processed selectively when and where they are required. In case of conventional sintering, all powders are processed simultaneously. 'Melting' refers to a particular case of powder processing in which powders are completely melted. The same process has also been named as laser cusing, direct metal laser sintering, laser generating, direct laser forming, direct laser fabrication but the name 'selective laser sintering/melting' is more widely used. The main difference between the two processes is that the former joins the powders by partial melting while the latter by full melting.

3.2.1 Why Selective Laser Sintering Is a Misnomer

The name 'selective laser sintering' is actually a misnomer, it erroneously implies that sintering is the main mechanism during selective laser sintering, while sintering has just come from tradition (powder metallurgy) and it actually never takes place during laser scanning of a powder bed. For sintering to take place, hours are required while scanning for a layer is completed in few seconds or minutes. Even in case of polymers it is the partial melting (Mokrane et al. 2018; Majewski et al. 2008) which is responsible for uniting materials; 'partial melting' means partial melting of a powder or partial melting of a powder bed or making a polymer molecule viscous. There is nothing less than this partial melting which will enable selective laser sintering to take place. It does not imply that there is no whatsoever relation between sintering and selective laser sintering. There is always sintering when there are circumstances for selective laser sintering to take place for hours, because sintering is a matter of hours and not a matter of seconds. Sintering takes place in a selective laser sintering setup when scanning takes place for hours for making a big part or for making a number of parts, or part is not removed from the hot machine for hours, or substrate temperature remains high for long. But, these sinterings do not define a shape; these sinterings are not responsible for providing complexity; these sinterings are not capable to do what sintering of selective laser sintering can do. The most appropriate name for selective laser sintering will be selective laser partial melting. However, the name sintering has been practiced because the name is used as a synonym for joining powders.

3.2.2 *Selective Laser Sintering*

In SLS, the main aim is to make a layer of predefined geometry by fusing powders using a laser beam. The process follows the following sequence: (1) a substrate is lowered down to a depth equal to layer thickness; (2) a powder layer is spread on the substrate; (3) the deposited powder layer is scanned by the laser beam to fuse powders at selected area. The sequence (1), (2), (3) is repeated until the desired fabrication is complete.

In an initial stage of the process, powders are placed in a powder container and are protruded from the container by an adjoining piston. Adjacent to the container, a scraper is placed which carries powders towards the substrate (or a build platform or inside a build chamber). The substrate is placed over a piston so that its vertical position can be changed by adjusting the piston. Scanning mirror is used to scan the deposited layer on the substrate using a laser beam coming from a laser source.

In step 1 of the sequence, the piston of the powder container moves upward and the piston of the substrate container moves downward. This step gives requisite powders to be carried away by the scraper and space on a substrate container for the powder to be deposited. In step 2 of the sequence, powders are deposited over the substrate and the position of the scraper changes to the right of the substrate. In the last step (step 3), deposited powders are scanned by a laser beam.

Above description shows basic necessities of the process: (1) formation of a powder bed, (2) consolidation of powders by a laser beam and (3) a mechanism to repeat above-listed first and second points. Deposited powder layer is termed as a powder bed. Point (1) has also given SLS a name 'powder bed process' or 'powder bed fusion'.

Instead of using a substrate over a build platform, the powder bed made directly over the platform could also act as a substrate. In this case, the final part is not needed to be cut off from the substrate as it is not attached to the solid substrate. Powder bed could also be formed by using a counter-rotating roller instead of a scrapper/blade, and the powder feeding could also be accomplished by using a hopper instead of a powder chamber. From the hopper, powders fall in front of the roller or scrapper, which then carries it away for deposition. The excess powder carried away falls into a trash that is equipped at the other side of the build chamber. Figure 3.2 is a schematic diagram of an SLS process.

The process generally occurs in a non-oxidative environment maintained by the presence of nitrogen or argon gas, while the temperature of the build chamber is

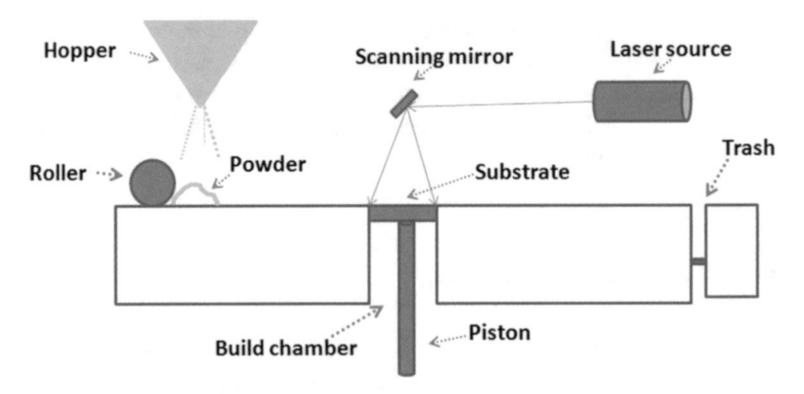

Fig. 3.2 A schematic diagram of SLS (Kumar 2014)

increased using attached heaters. The powder feeding, scanning, temperature, atmosphere, build and deposition system are computer-controlled.

3.2.3 Selective Laser Melting

SLM refers to a case in which full melting of powders occurs. Though, for polymer, even in the case of the full melting the name SLS instead of SLM is used. It is because fully melted polymer parts remain always porous. Henceforth, it is more logical to use SLS instead of SLM for polymers as SLM is synonymous with providing almost dense metal parts.

When metal or ceramic powders are fully melted, then they come under SLM. The description of SLM is the same as that of SLS, and the above description of SLS applies to SLM.

Roles of various parameters such as laser power, types of laser, laser mode, laser spot size, powder size distribution, powder size, powder flowability as well as various scanning and building strategy are given elsewhere (Kumar 2014).

3.3 Process Parameters

In this section, various parameters related to the process are described and their effects on properties are explained. Figure 3.3 shows schematic diagrams of processing of powder layer with a laser beam. While Fig. 3.3a illustrates layerwise processing by the laser beam, Fig. 3.3b illustrates clearly scan spacing and beam overlap.

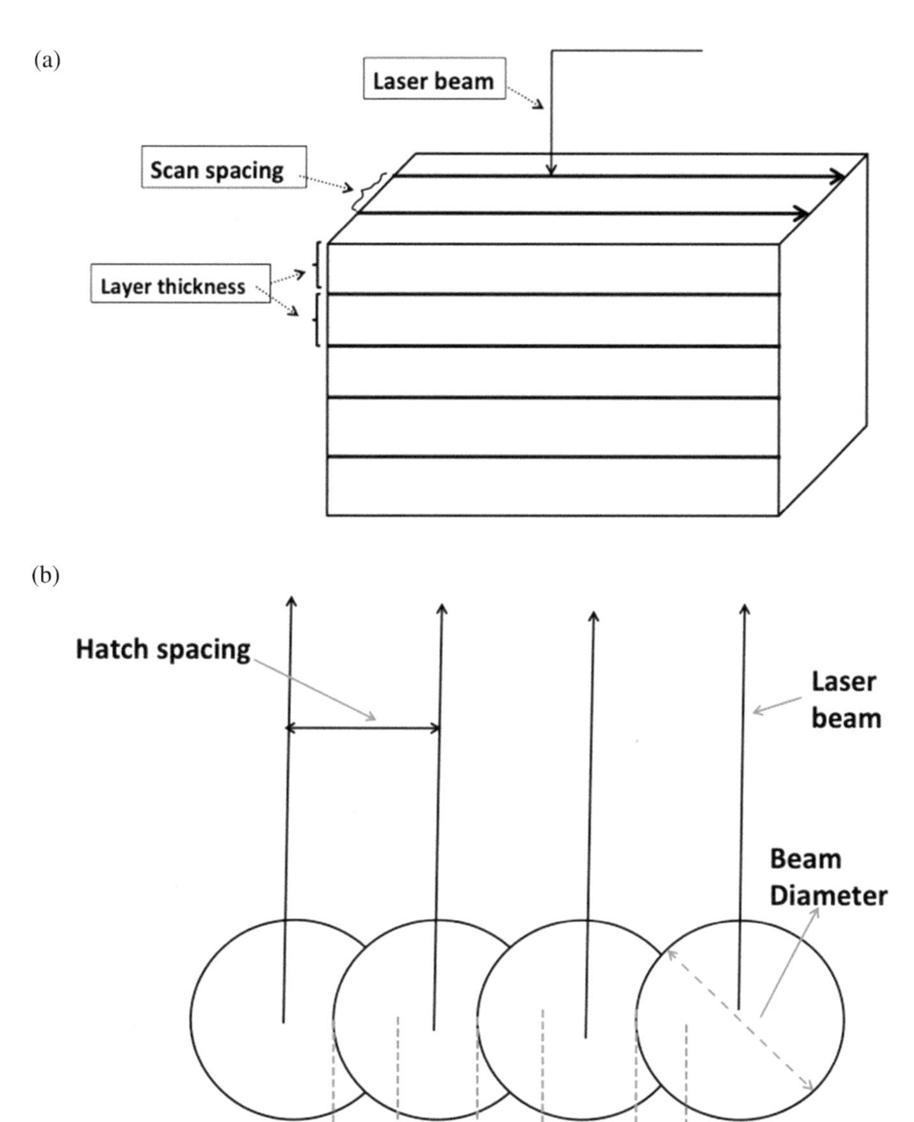

Fig. 3.3 Diagrams showing (**a**) layerwise processing by laser beam, (**b**) scan spacing and beam overlap (Kumar 2014)

3.3.1 Scan Spacing

Scan spacing is a separation between two consecutive laser beams. It is also called hatch spacing or hatch distance. As shown in Fig. 3.3b, it is measured by a distance from the centre of one beam to the centre of the next beam. Scan spacing is directly proportional to the production speed. If it is high, it will take less time for a laser to scan the layer, while if it is low, a number of scanning needs to be executed to process the whole layer. Smaller scan spacing is required for making thin features.

In order to have a large scan spacing, a large laser spot size is required. Otherwise, there remains a gap between two consecutive scans resulting in porous parts. For processing with a larger spot size, higher laser power is required to supply necessary laser energy. It implies that in a given LPBF system, the maximum scan spacing obtained is limited. In order to avoid any porosity formation at the boundaries of scans, some overlap, as shown in Fig. 3.3b, is made. Overlap is necessary because in a typical Gaussian beam, laser power at the centre of the scan is higher than at the boundary of the scan resulting into melting at the centre while heating at the boundary. Creating the overlap compensates this less heat generation at the boundary.

3.3.2 Scan Speed

Scan speed is the rate at which a laser beam scans a line on a powder bed. With an increase in scan speeds, production speeds increase. Scan speed and laser energy density are related as follows:

$$E_v = \frac{P}{S_d \times V_s} \tag{3.1}$$

where, E_v = laser energy density, P = laser power, V_s = scan speed, S_d = spot size.

Equation (3.1) shows that at a high scan speed, laser energy density is low and may not be sufficient to process the powder bed. This could be compensated by increasing the laser power. However, at a very high scan speed and laser power, the time is not sufficient for heat to diffuse across the whole powder bed; it could lead to insufficient melting and ablation of the powder. Therefore, the above equation holds good only within a limit which is determined by the types of materials and other process conditions such as temperature and pressure of the environment. The value of the scan speed used is in the range of 0.1 ms^{-1} and 15 ms^{-1}.

The scan speed in SLM determines melt pool length. Higher scan speed gives rise to longer and thinner melt pools which have higher chances to break into several smaller melt pools (ball) due to Rayleigh instability.

3.3.3 Layer Thickness

Layer thickness is the thickness of a slice of a 3D CAD model of a part which is transformed into a physical layer by laser processing as shown in Fig. 3.3. It is the same as a powder layer used in the process which is set by changing the height of a build platform. Layer thickness is another important parameter directly related to the production speed. With an increase in the layer thickness, higher production speed is achieved, while with its decrease, higher precision is achieved.

Higher laser energy is required for processing thicker layers. However, there is a limit to which laser energy could be increased because supply of high energy sometimes causes distortions on the surface and gives rise to inaccuracies. This could be avoided by scanning the same surface twice with lower energies.

Laser energy density is given by the following relation:

$$E_v = \frac{P}{S_s \times V_s \times L_t} \tag{3.2}$$

where, E_v = laser energy density, P = laser power, V_s = scan speed, L_t = layer thickness, S_s = scan spacing.

Equation (3.2) gives the laser energy density across the thickness of the powder bed. It is to be noted that Eq. (3.1) gave the energy density only on the surface of the powder bed. Equation (3.2) holds when laser spot size is always bigger than scan spacing. If laser spot size is smaller than scan spacing, then Eq. (3.2) will consist of term spot size instead of scan spacing.

Thinner layer needs low energy density but furnishes dense parts with low surface roughness. For making a thin layer, small powder size is required. Using thin layers not only increases the production time but also the production cost. Lower layer thickness also means lower shrinkage after melting by moving laser beam, which will increase the dimensional accuracy and surface smoothness.

Selection of thickness of layers depends also upon the geometry of the part to be fabricated. When a curved object is fabricated layerwise, then layer being rectangular in nature does not coincide with the contour of the curved object, it leads to a gap on the side of the object as shown in Fig. 3.4. This is also called staircase effect. The size of the gap depends upon the layer thickness. For thinner layers, gap is smaller. But the gap remains always present in the layerwise built; the effort is to minimize this gap so that the resulting contour will be acceptable.

In order to maximize the production speed without losing the precision due to the staircase effect, the thickness of layers in a given built is optimized. For a vertical

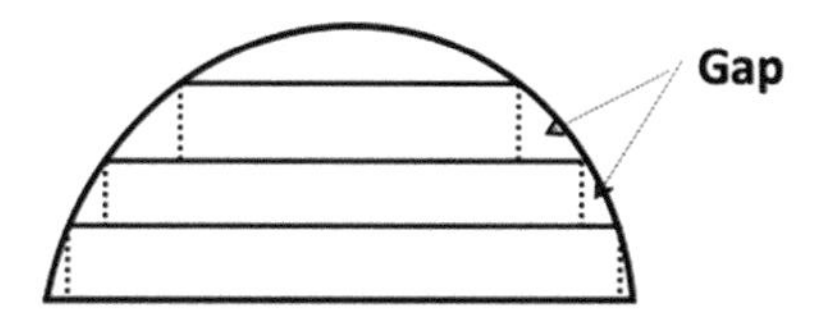

Fig. 3.4 Staircase effect causing gap between two layers in a curved part (Kumar 2014)

edge section of the part, higher layer thickness is selected while for a slope edge section, smaller layer thickness is selected.

The effect of both spot size (beam diameter) and hatch spacing together is given by Eq. (3.3). Previously used Eq. (3.2) is a special case of Eq. (3.3) when spot size equals scan spacing (Kumar and Czekanski 2017):

$$\begin{aligned} E_v &= \frac{P}{V_s \times S_d \times L_t} \times \left[1 + \frac{(S_d - S_s)}{S_d} \right] \\ E_v &= \frac{P}{V_s \times L_t} \times \left[\frac{(2S_d - S_s)}{S_d^{\,2}} \right] \end{aligned} \tag{3.3}$$

where, E_v = laser energy density, P = laser power, V_s = scan speed, S_d = spot size, L_t = layer thickness, S_s = scan spacing.

3.4 Why There Are No Four Binding Mechanisms

There are four binding or fusion mechanisms mentioned in laser powder bed fusion (LPBF) (Kumar 2014; Gibson et al. 2010). It is imperative to know what are the main reasons responsible for joining powders in LPBF: this implies the very cause without which it will not be possible for powders to convert into a 3D shape. The process starts when powders are placed in an LPBF system, and the process completes when a processed material is taken out. There are two mechanisms which work during this time and without which conversion into a 3D shape will not happen. These are (1) liquid-phase sintering (LPS) or partial melting and (2) full melting (FM). When the process ends, that is the processed material (part) is taken out from the machine, whatever further processing is done on the part will not add to the process which is already ended. In other words, at this stage improving the part is not a synonym for improving the process. During this post-processing time, the part will be further improved; the mechanism invoked earlier that is LPS and FM, does not help to explain this improvement; one more mechanism, that is solid state sintering (SSS), is used to explain this improvement. Thus, there are two binding mechanisms (LPS, FM) which work during LPBF, and there are three mechanisms (LPS, FM, SSS) which work for LPBF parts. One more mechanism, that is chemical-induced binding (CIB), is mentioned, which could be a possible mechanism for future. Four mechanisms are given below and it is described why SSS and CIB are not binding mechanisms.

3.4.1 *Liquid Phase Sintering*

Liquid-phase sintering (LPS) is a widely used mechanism in SLS. During laser-material interactions, some of the powders are converted to liquid; the liquid thus generated flows and fills up the pores made by adjacent powders causing the powders to be joined. Consequently, in order to join majority of the powders, melting some of the powders rather than all powder is a sufficient condition.

This mechanism is also referred to as partial melting because for joining powders part of a single powder or part of a bunch of powders are melted. Partial melting is considered as a case of full melting in which either laser energy is not sufficient to cause complete melting of powders or the laser energy is intentionally set not sufficient (by adjusting experimental parameters) to cause complete melting.

There are various ways powders are engineered to facilitate the consolidation by this mechanism:

- **Coating of a powder**: Powders are coated with a low melting point material (either polymer or metal) so that during laser treatment the coating will preferentially melt which will act as a binder to bind unmelted cores. Examples are steel coated with polymer, iron coated with copper, sand coated with phenol etc.
- **Powder mixture**: Low melting point material is mixed with a high melting point material so that during laser treatment, low melting point materials will melt and join the high melting point materials. In this type, the former material is called binder while the latter is called a structural material. Examples are WC + Co (Kumar 2018), SiC + PA, stainless steel + polymer. In these examples, WC, SiC and stainless steel act as structural materials while Co, PA and polymer act as binder materials in their respective mixture.
- In another type of mixing, same material having two sizes (small and big) are mixed. Small size powder preferentially melts because of its high surface-to-volume ratio and acts as a binder while a big size powder acts as a structural material.
- **Composite powder**: Powder mixture may be segregated during powder bed deposition giving rise to non-homogeneous melting of powders responsible for non-uniformity in properties. This problem is avoided by mechanically alloying different powders instead of just mixing them and making composite powders.

The mechanism economizes laser energy required for consolidation; it is done by engineering powder surface, composition and mixture. This is also accomplished by adjusting process parameters so that only that amount of melting will take place which is sufficient enough to result appreciable (pore- and defect free) consolidation – this will additionally give rise to a lower surface roughness (Kumar 2014).

3.4.2 Full Melting

In this mechanism, powders are completely melted by a scanning laser beam which implies that this mechanism in comparison to LPS gives advantages – to be free from finding parameters for economizing or minimizing laser energy, to be free from seeking prior information about types and amount of binder materials in a powder mixture. Complete melting gives rise to a melt pool which on solidification forms a solidified melt pool – a unit which adds on with a progress in laser scanning – finally giving a desired part. In the course of formation of a melt pool, powders melt and disappear, there are no more powders left in the melt pool; therefore, in this mechanism unlike other mechanisms (LPS, SSS), the problem to bind powders does not exist. Thus, the problem in this mechanism is not how to bind powders but how not to bind with other powders. Here, other powders connote those powders present adjacent to melt pools, underneath the melt pool in case of overhangs or in cases (specially in polymers) as a powder bed in place of a substrate. The binding of the melt pool with other powders depends upon the size, viscosity, surface tension and dwell time of the melt pool as well as upon the capillary determined by packing, size and size distribution of powders. Binding of the powders adjacent to the melt pool happens but is not required because the effect of the melt pool on powders (such as partial melting or joining) is superseded by subsequent pass of the laser beam; at the boundary of the geometry of the intended part, this type of interaction of the melt pool with powders increases surface roughness (Bian et al. 2018) (of the side surface) resulting in high difference in the surface roughness of side surface and top surface of any built (Mumtaj and Hopkinson 2009); if the binding is more, it will cost the accuracy of the geometry; less the binding, better it is. Therefore, this mechanism of binding is governed by having control and capability to prevent such binding – in this mechanism of binding it is required to have a mechanism of no-binding to prevent such undesirable interactions. In case of melt pool of overhangs facing powders underneath or in case of melt pool of the first layer of the built (fabricated on a powder bed instead on a substrate) facing powders underneath, minimization of binding of powders with melt pools furnish smooth surface (Bian et al. 2018).

This mechanism provides high density, high strength, variable strength, a range of ductility, no porosity (except isolated porosity in some cases), better microstructure than wrought or cast materials, novel microstructure not possible through conventional manufacturing and high rate of production. The high rate is achieved by employing higher scan speed in comparison to LPS.

3.4.3 Why Solid State Sintering Is Not a Binding Mechanism

Solid state sintering (SSS) binds powders by diffusion of atoms. The diffusion could be of types – volume diffusion, surface diffusion or boundary diffusion. The diffusion gives rise to the formation of neck at the boundary of two powders; the neck gets extended which fills the space with ongoing diffusion and binds the powders. For the diffusion to take place powders need to be heated to a certain temperature, while for the diffusion to be completed sufficient time (from several minutes to hours) is required. The temperature should not be more than the melting point of powders otherwise the powders would no longer be in a solid state.

It has been shown that by heating titanium powders with laser beam for sufficiently long time (5 s), they could be joined (Tolochko et al. 2000). But in laser powder bed fusion (LPBF), laser-material interaction takes shorter interval (of the order of ms or lower depending upon the scan speed), which is not enough to consolidate any materials by this mechanism. Though, the mechanism is used during post-SLS furnace treatment (sintering, debinding, infiltration) when the porous parts need to be densified and strengthened. Furnace treatment allows sufficient time and temperature required for this mechanism to occur.

During LPBF, process chamber is heated either to aid processing or to decrease thermal stress; a hot process chamber combined with longer processing time induce significant diffusion to take place. Due to the diffusion, joining among powders takes place everywhere within the chamber which is a significant demerit of the process as it decreases the recyclability of the powders, increases the time to clean the parts as powders get attached to them and increases surface roughness of the part. For a part that is going to be built by partial melting or liquid phase sintering (LPS), this diffusion has some impact as it will give rise to solid state sintering (SSS) and some weakness of the part is being alleviated. But, SSS is not a mechanism by which the part is taking shape, and the contour of the part is not determined by SSS but by some other mechanisms (LPS). In the absence of LPS and in the presence of SSS, there will not be any LPBF part; therefore, SSS is not a binding mechanism for LPBF. It can be argued that the part is still somewhat strengthened by SSS, then why not should it be considered at least a secondary binding mechanism for LPBF. The following are the reasons:

1. SSS is not aimed at. It is a by-product or waste product of heating the process chamber for some other purposes. The chamber is heated or kept at elevated temperature for aiding the processing – so that low laser power instead of high laser power is required to complete the process – thus the problem associated with application of high laser power is avoided; or the chamber is heated to decrease the thermal gradient so that crack formation can be avoided. SSS is not aimed at because it is not required, if it will be required then it can be executed during post-processing stage; however, during this stage the crack cannot be undone, and the already completed process cannot be attended to retrospectively. SSS is also not aimed at because it has not a shaping capability, it cannot make contours. SSS is size dependent; if a milimetre height part is formed, SSS is not significant while if a half metre height part is formed, SSS is over-significant.
2. The advantage got due to diffusion-induced strengthening is counter-balanced by the disadvantage it gave, such as loss of surface smoothness and problem during cleaning. If for getting more advantage, diffusion is planned to last longer, it will

be difficult to retrieve the part from sintered powders. Therefore, this advantage is not an advantage in true sense.

If the mechanism LPS in above sentences is replaced with FM in case of full melting by laser beam, the impact of diffusion will be more insignificant – from minimal to zero.

3.4.4 Why Chemical-Induced Binding Is Not a Binding Mechanism

If some material is found or engineered in the future which during laser-material interaction gives rise to some compounds which will be responsible for binding remaining powders, then this mechanism for binding will come under chemical-induced binding (CIB) provided the compound is not in the form of liquid; if it is in the form of liquid then it would not be possible to distinguish this mechanism from LPS. Another condition is that the formation of the compound is the only reason due to which powders bind together; there should not be any side effect of laser-material interaction such as some melting of some isolated powders which also contributes to binding. There is not a single incidence yet, as published, which satisfies all conditions and come under CIB.

If laser-material interaction results in a chemical compound accompanied by liquid, the liquid is responsible for binding; depending on the amount of liquid, it will create a condition of either partial melting or full melting. In case of an exothermic reaction, the chemical compound will be accompanied by disproportionately high amount of the liquid, but still the liquid is responsible for binding and not the exothermic energy on its own. Binding with an involvement of liquid is already covered under mechanisms LPS or FM irrespective of the cause of the creation of the liquid. In this case, the cause is chemical, and there is no reason why this case should be deprived of its right place, that is LPS or FM. If a cause would be the reason for naming a mechanism then LPS or FM would not be sufficient but can be expanded such as coating-induced binding, temperature-induced binding, pressure-induced binding, gas-induced binding, composite-induced binding, powder size-induced binding etc.

3.5 Methods for Increasing Fabrication Rate

Increasing fabrication rate has a direct consequence on decreasing the product cost and increasing the production efficiency (Thompson et al. 2016). There are a number of methods by which high fabrication rate can be achieved in a powder bed fusion system. These methods can also work with other processes These methods are as follows:

3.5.1 *Increasing Layer Thickness*

Increasing layer thickness will decrease the number of layers required to make a part (Shi et al. 2016), thus the number of times powder will be deposited to make a layer will decrease, which in turn will decrease the total fabrication time and will hence increase the fabrication rate. However, this option for increasing the fabrication rate has a limitation – increasing it more than a certain thickness will not let the effect of melting to reach to the base of the layer – powders will either remain unaffected or will bind by other mechanism (by solid state sintering if build continues for long). Thus, there is a limitation to which the fabrication rate can be enhanced by this method (Gao et al. 2015). For certain geometries, increasing layer thickness will also increase staircase effect and decrease accuracy (details given in Chap. 10). This method is applicable in all layer based AM processes (Lim et al. 2012).

3.5.2 *Dividing the Processing Area*

Dividing the processing area in a number of zones and processing each zone simultaneously using a number of laser beams is another method to increase the fabrication rate. A single layer is processed by a number of beams parallelly (Fig. 3.5a), fabrication time is equal to the total area divided by the number of laser beams when

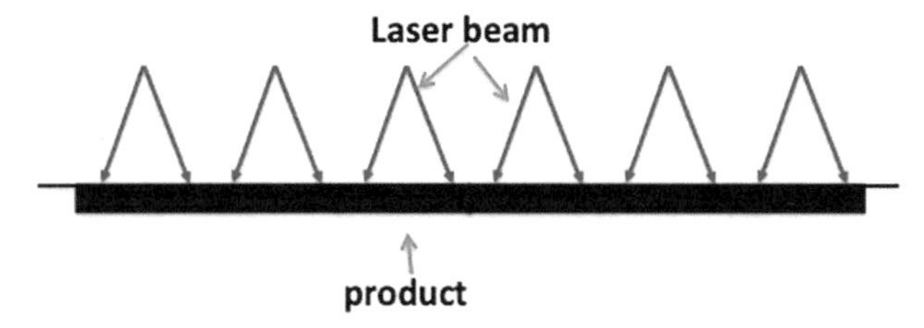

(a) Application of a number of laser beams for making a single part

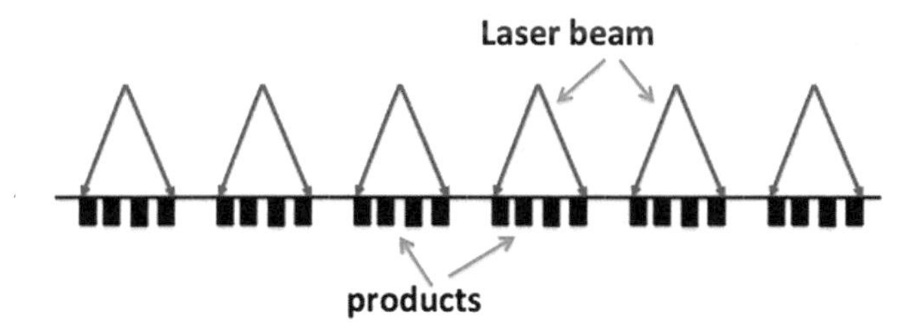

(b) Application of a number of laser beams for making a number of parts

Fig. 3.5 Application of a number of laser beams for making (**a**) a single part, (**b**) a number of parts

each laser beam processes an equal area. Fabrication time thus decreases with an increase in the number of beams and thus the fabrication rate increases. In this option, the reason for an increase of the rate is not exclusive to this mechanism or the process – it is a general case of addition of efforts (in this case, addition of efforts for addition of materials). This option is limited by: logistics – complex algorithm for sequence of operation with many beams, size of the machine; sustainability – if one laser beam fails then it needs to be fixed before using all beams: cost effectiveness and logic – why not making many machines equipped with a single laser each than many lasers in a single machine, it will allow to work with different materials in separate machines.

This method has limitation for making a small part where size of the boundary between two zones is not significantly smaller than the area of the zones, when two zones are processed by two separate laser beams. Different sizes of melt pools will be formed by different laser beams having small variation or different levels of noise. In case of a bigger part, this small difference will be insignificant, but in case of a small part, this small difference will not be small enough. Nevertheless, in case of making a huge number of smaller parts (e.g. mass customization) in a big PBF machine equipped with the capability to employ parallel scanning enabled by a number of laser beams (Fig. 3.5b), the task of processing (or producing) such huge numbers will be shared equally by each laser beam – addition of efforts; it will increase overall fabrication rate but it will neither increase the fabrication rate of a single part nor will require synchronization of many laser beams working on a single part. The single part thus fabricated in a multi-laser setting will not be different from a part if fabricated in a similar setting but using just a single laser.

Figure 3.5a shows laser beams coming from six laser sources (red colour) contribute to make one part (black colour) while Fig. 3.5b shows laser beams are coming from six laser sources (red colour), they each scan equal space (1/sixth of space) and make four parts (black colour) each.

3.5.3 *Optimizing the Process as per the Need*

If high strength is not required for all sections of a part, the section which does not require high strength will be scanned with fast scan speed- this will increase fabrication rate (Kniepkamp et al. 2018). This section can also be scanned by decreasing the overlap between two adjacent scans, it will decrease the number of scans required to cover a section – it will decrease the time spent in scanning and will thus increase the fabrication rate in these sections. Hence, optimizing the process parameters as per the requirement of different sections of a part will not only pave the way for increasing the fabrication rate but also decreases the energy input.

3.5.4 *Changing the Orientation of Design to Be Fabricated*

A part can be made in many ways by changing its orientation (changing the angle between one of its features and build direction) (Zhang et al. 2017), if changing the orientation will result in a decrease in number of layers required for completion of the built, fabrication time will be lowered. However, this option for increasing the fabrication rate has limitations such as decrease in mechanical properties (heat accumulated on top layer will not be same in both orientations), need for a support structure (Calignano 2014), change in microstructure size (the length of columnar grain will not be same in both orientations) and orientation affecting the properties (Barba et al. 2020), decrease in surface roughness and geometrical accuracy, in the worst case it is impossible to build for even a small alteration in the orientation. Figure 3.6 shows an example of the effect of orientation; building in an orientation demonstrated in Fig. 3.6a will increase the fabrication rate because of a decrease in height of the build but will have to create a big overhang which is not supported on a solid base. Changing the orientation by 90° as shown in Fig. 3.6b will increase the height of the build and thus increase the fabrication time but it will decrease the size of the overhang required to be built and thus will be free from difficulties of making bigger overhang. Consequently, former orientation (Fig. 3.6a) furnishes higher fabrication rate or higher build rate but having lower surface finish while the latter orientation (Fig. 3.6b) furnishes lower fabrication rate but higher surface finish. This example shows that changing the orientation means making a choice, and it is not uncommon to have a number of examples free from such limitations.

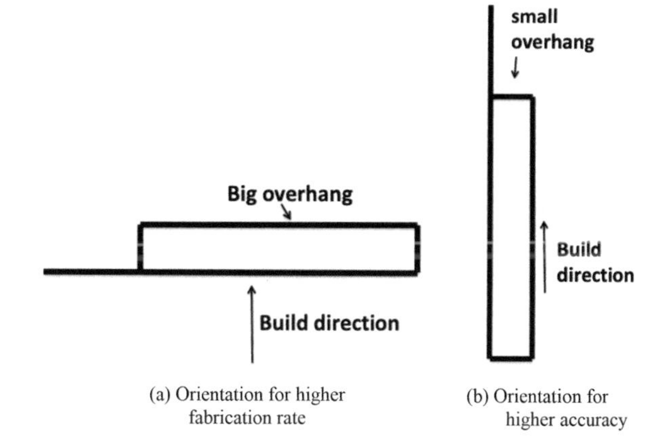

Fig. 3.6 Different orientation of the same design of a part for different needs: (**a**) orientation for getting higher fabrication rate, (**b**) orientation for getting higher geometrical accuracy

3.5.5 Increasing Scan Speed

The dominant factor among all factors which affect fabrication rate is the scan speed; it has direct positive consequence on the rate as with an increase in scan speed, fabrication rate increases. Increasing the speed will also increase the rate by exercising other above-mentioned options.

The highest scan speed that could be reached is limited by following:

1. Evaporation of powders rather than melting: at high speed of a high-power laser beam, top section of a powder accumulates high amount of heat leading to evaporation.
2. Formation of a longer melt pool (at high scan speed) which breaks into shorter melt pool giving rise to discontinuity: it is due to Rayleigh instability which states that if the length by width ratio is more than π (a critical value depends upon interface) then it will break. Longer pool will no longer be a longer pool if it solidifies on the way – if solidification rate is faster than or equal to the scan speed, there will never be a longer pool {in case of aluminium alloy scanned at 1 m/s, the solidification rate is almost the same (Tang 2017)}. Cooling rate remains always high in an experimental setup working on the basis of this mechanism; this is one of the reasons why Rayleigh instability does remain not even one among many problems for consolidation. Another reason is that experiments are not conducted at higher scan speed – it is better to have no-defect products at low scan speed than to have any products at high scan speed. A scan speed can be termed low or high depending upon materials and experiments. A low scan speed of 689 mm/s is considered as a high scan speed when it gives porous parts (Gong et al. 2014).

Which one of the above-mentioned reasons will be the main reason for failure to make defect-free parts at high scan speeds (100 m/s or more); at a certain highest speed, only evaporation will take place – there will be no melt pool and there is no question of the type of melt pools; consequently, the first reason is the cause of failure. Increasing the pressure of processing chamber will increase this speed as the pressure will suppress the evaporation – more pressure means higher speed, there comes a limit beyond which application of pressure does not work – either due to its effect on the melt pool such as depression or due to safety aspect involved with working at high temperature and high pressure. Increasing the speed further without increasing the pressure will bring forth a result no more different than evaporation unless laser power is not increased proportionally; in the absence of increased laser power, there will be neither evaporation nor melting but just heating of powder bed. With a decrease in speed, there will be a certain speed at which evaporation will stop and melting will start, if solidification rate will be similar to the speed, this mechanism will be able to furnish a part. If the solidification rate will be lower, two cases will happen: (a) part formation will fail due to Rayleigh instability and (b) part formation will not fail if some measures are taken. One measure – such as increasing the cooling rate by decreasing the temperature of the substrate – will increase the solidification rate but part may still fail due to the formation of cracks. Another measure is to employ laser pulse (of the order of ms in order to prevent evaporation associated with ultrashort laser pulse) instead of continuous laser beam – it will create smaller pool. Applying the mechanism at higher speed will leave few options to exercise for making a defect-free part, thus decreasing the freedom to vary microstructures and properties.

3. High speed requires high laser power to maintain required laser energy density for melting (Sun et al. 2016). High laser power means higher cost; it will increase the cost of the machine. Working with high laser power means an extra need to manage high heat generated. Management of high heat means upgrading the device (laser, mirror etc.) and incorporating new cooling device lest the machine or part of the machine be damaged. Upgrading the device depends upon the availability of a device – a scanner working at high power and high speed may not be available, and if available, may not be expected to last longer. However, in case of EBM, there will not be any damage to scanning device but the lifetime of electrodes can be lowered at constant generation of high power.

3.5.6 *Adopting Linewise and Areawise Scanning*

Pointwise scanning means scanning using a point source such as electron beam, a laser beam or a jet from a single nozzle – in these cases, diameter of an electron beam at the powder bed is the size of a point source, laser spot size of the laser beam is the size of a point source, size of the drop on the surface from a nozzle is the size of a point source in binder jetting. Using such point source in pointwise scanning takes time for processing a layer – point source has to process one point on the powder bed then it has to move to another point – processing moves from point to point, work is done bit-by-bit; in case of faster scanning processing moves from point to point faster. Though, it moves faster, but the problem is it has to move from point to point, and a powder bed has a myriad of points, and it has to do a lot of work. The above methods for increasing fabrication rate, which relies on pointwise scanning, suffer from inherent limitation of pointwise scanning.

Changing from pointwise scanning to linewise scanning (Thomas et al. 2006; Sillani et al. 2019) or to areawise scanning is another method to increase the fabrication rate. In linewise scanning, a number of laser beam sources (diode emitters) will be fixed together in a horizontal line, and the line will move to process layers, there is no role of scanners (to deflect laser beams) (Arredondo et al. 2017; Dallarosa et al. 2016). In areawise scanning, all heat point sources will be fixed in area over a powder bed and they will process the bed simultaneously without any either movement of heat source or deflection of heat radiation (Holt et al. 2018) (given in Chap. 5).

3.6 Repair in Laser Powder Bed Fusion

Repair is a cost-effective measure taken to save either some or all – scarce resources, energy, time and labour of original manufacturing; it has been used in some AM processes such as laser engineered net shaping (LENS) (Onuike and Bandyopadhyay 2019) and cold spray additive manufacturing (CSAM) (Li et al. 2018; Huang et al. 2019). However, it is not generally used in PBF; in some cases when a part is made from a customized expensive material and the damage is not severe, repair can be employed.

For repair to be carried out, the following two conditions must be met:

1. Damage of a part should be at such a location of the part that the damaged surface could be kept parallel to the platform or build plane, it will ensure no collision with roller or scraper during building of the surface. For example, in Fig. 3.7a if all features, 1, 2, 3, 4 and 5, get snapped from rectangular block ABCD then the damaged surface AB can be kept parallel to the build plan and can be repaired; Fig. 3.7a shows front view of rectangular block ABCD and surface AB by rectangle ABCD and line AB, respectively.
2. The part could be fixed on a platform such that the damaged surface is parallel to the platform. For example, lower surface CD (Fig. 3.7a) of rectangular block ABCD is plane without having any feature with complex geometry and is suitable to be fixed on the platform keeping the surface AB parallel to the platform (Fig. 3.8a). If the lower surface CD is tilted as in Fig. 3.9a or the lower surface has a long thin feature as in Fig. 3.9b, then the part may not be fitted well on the platform for further repair to be performed.

For a damaged part to be repaired, it needs to follow certain steps for repair. Figure 3.7a shows an original part having five features while Fig. 3.7b shows the same part having one feature numbered 3 is missing; features are blocks of equal height having certain cross-section. In order to repair the damaged part (Fig. 3.7b), the missing feature should be regrown at the same location. It is not possible in PBF to regrow the feature selectively unless in exceptional cases, when the height of the

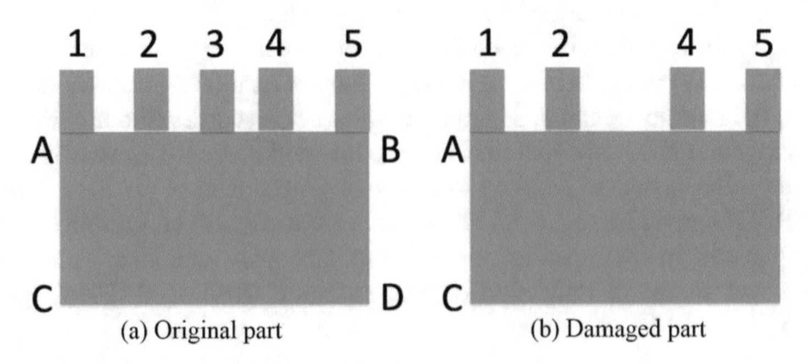

Fig. 3.7 Schematic diagram of an original part and a damaged part: (**a**) original part, (**b**) damaged part.

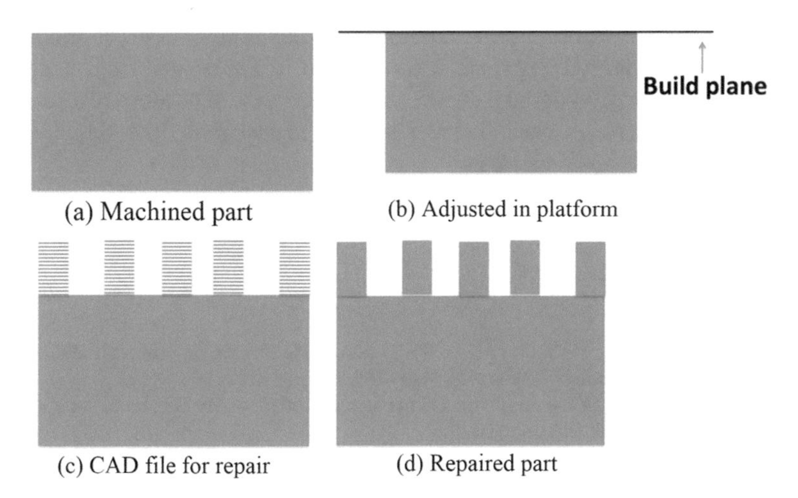

Fig. 3.8 Various stages of a repair: (**a**) machined part, (**b**) adjusted in platform, (**c**) CAD file for repair, (d) repaired part.

feature is approximately similar to the layer thickness. Therefore, all features should be removed by machining to build all new features instead of just one new feature. Figure 3.8a shows the machined part without any features which can be fitted well in the platform (Fig. 3.8b) because surface AB is parallel to the build plane; lower parallel surface CD will help it get fixed. A CAD file as shown in Fig. 3.8c having all features sliced is used to build features in usual layer by layer to make final part (Fig. 3.8d).

Need to remove four features for repairing just one feature brings a question mark on the efficiency of PBF for repair, but economically and resource-wise it is certainly better than non-using the damaged part and opting for a new part.

If the original part is damaged severely such as shown in Fig. 3.9c, where there is no missing feature but a triangular cut is reached inside the bulk of the part, repair may not be an option. If it is machined parallel to AB as done earlier (Fig. 3.8a), approximately three-fourth of the part will be machined out; if it is machined parallel to BC so that repair could be done by changing the orientation, approximately

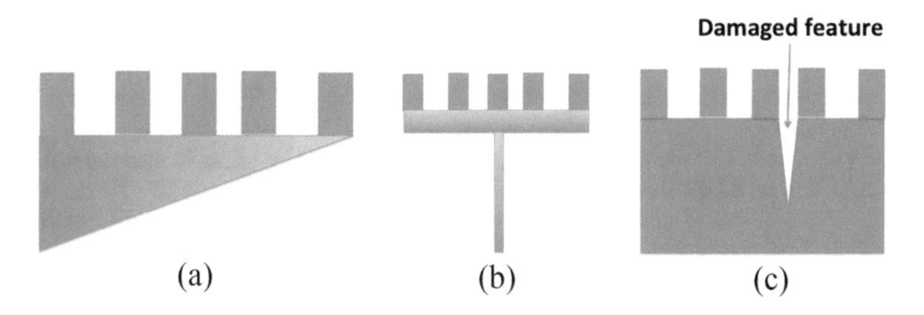

Fig. 3.9 Non-repairable part in a PBF system: (**a**) oblique lower surface, (**b**) thin feature in lower surface, (**c**) deep damaged feature

two-fifth of the part will be removed; if it is machined from side AC, approximately three-fifth of the part will go; there is no option if it is machined from side CD as approximately four-fifth of the part will be removed as well as all five features will be separated. In all cases, machined part is not large enough to drive significant difference in repairing and fabricating.

References

Arredondo MZ, Boone N, Willmott J et al (2017) Laser diode area melting for high speed additive manufacturing of metallic components. Mater Des 117:305–315

Barba D, Alabort C, Tang YT et al (2020) On the size and orientation effect in additive manufactured Ti-6Al-4V. Mater Des 186:108235

Baumers M, Tuck C, Hague R (2015) Selective heat sintering versus laser sintering: comparison of deposition rate, process energy consumption and cost performance. In: SFF proceedings, pp 109–121

Bian Q, Tang X, Dai R, Zeng M (2018) Evolution phenomena and surface shrink of the melt pool in an additive manufacturing process under magnetic field. Int J Heat Mass Transf 123:760–775

Brown R, Morgan C T, Majweski C E (2018) Not just nylon—improving the range of materials for high speed sintering. In: SFF proceedings, pp 1487–1498

Calignano F (2014) Design optimization of supports for overhanging structures in aluminum and titanium alloys by selective laser melting. Mater Des 64:203–213

Dallarosa J, O'neill W, Sparkes M, Payne A (2016) Multiple beam additive manufacturing. Patent WO2016201309A1

Enneti RK, Prough KC, Wolfe TA et al (2018) Sintering of WC-12%Co processed by binder jet 3D printing (BJ3DP) technology. Int J Refract Met Hard Mater 71:28–35

Gao W, Zhang Y, Ramanujan D et al (2015) The status, challenges, and future of additive manufacturing in engineering. Comput Aided Des 69:65–89

Gibson I, Rosen DW, Stucker B (2010) Additive manufacturing technologies: rapid prototyping to direct digital manufacturing. Springer, New York

Gong X, Lydon J, Cooper K, Chou K (2014) Beam speed effects on Ti–6Al–4V microstructures in electron beam additive manufacturing. J Mater Res 29(17):1951–1959

Holt N, Horn AV, Montazeri M, Zhou W (2018) Microheater array powder sintering: a novel additive manufacturing process. J Manuf Process 31:536–551

Huang CJ, Wu HJ, Xie YC et al (2019) Advanced brass-based composites via cold-spray additive-manufacturing and its potential in component repairing. Surf Coat Technol 371:211–223

Jerby E, Meir Y, Salzberg A et al (2015) Incremental metal-powder solidification by localized microwave-heating and its potential for additive manufacturing. Addit Manuf 6:53–66

Khoshnevis B, Zhang J, Fateri M, Xiao Z (2014) Ceramics 3D printing by selective inhibition sintering. In: SFF proceedings, pp 163–169

Kniepkamp M, Harbig J, Seyfert C, Abele E (2018) Towards high build rates: combining different layer thicknesses within one part is selective laser melting. In: SFF symposium proceedings, pp 2286–2296

Korner C (2016) Additive manufacturing of metallic components by selective electron beam melting- a review. Int Mater Rev 61(5):361–377

Kumar S (2010) Selective laser sintering: recent advances. Pacific International Conference on Applications of Lasers & Optics, Wuhan, China, March 23–25

Kumar S (2014) Selective laser sintering/melting. Compr Mater Process 10:93–134. Elsevier Ltd

Kumar S (2018) Process chain development for additive manufacturing of cemented carbides. J Manuf Process 34:121–130

Kumar S, Czekanski A (2017) Optimization of parameters for SLS of WC-Co. Rapid Prototyp J 23(6):1202–1211

Li W, Yang K, Yin S et al (2018) Solid-state additive manufacturing and repairing by cold spraying: a review. J Mater Sci Technol 34(3):440–457

Lim S, Buswell RA, Le TT et al (2012) Developments in construction-scale additive manufacturing processes. Autom Constr 21:262–268

Liravi F, Vlasea M (2018) Powder bed binder jetting additive manufacturing of silicone structures. Addit Manuf 21:112–124

Majewski C, Zarringhalam H, Hopkinson N (2008) Effect of the degree of particle melt on mechanical properties in selective laser-sintered Nylon-12 parts. Proc Inst Mech Eng B J Eng Manuf 222(9):1055–1064

Mokrane A, Boutaous M, Xin S (2018) Process of selective laser sintering of polymer powders: modeling, simulation and validation. CR Mec 346:1087–1103

Mumtaj K, Hopkinson N (2009) Top surface and side roughness of inconel 625 parts processed using selective laser melting. Rapid Prototyp J 15(2):96–103

Onuike B, Bandyopadhyay A (2019) Additive manufacturing in repair: influence of processing parameters on properties of Inconel 718. Mater Lett 252:256–259

Shi X, Ma S, Liu C et al (2016) Performance of high layer thickness in selective laser melting of Ti6Al4V. Materials 9:E975

Sillani F, Kleijnen RG, Vetterli M et al (2019) Selective laser sintering and multi jet fusion: process-induced modification of the raw materials and analyses of parts performance. Addit Manuf 27:32–41

Sun Z, Tan X, Tor SB, Yeong WY (2016) Selective laser melting of stainless steel 316L with low porosity and high build rates. Mater Des 104:197–204

Tang M (2017) Inclusions, porosity and fatigue of AlSi10Mg parts produced by selective laser melting. PhD thesis, Carnegie Mellon University

Thomas H R, Hopkinson N, Erasenthiran P (2006) High speed sintering- continuing research into a new rapid manufacturing process. In: SFF proceedings, pp 682–691

Thompson MK, Moroni G, Vaneker T et al (2016) Design for additive manufacturing: trends, opportunities, considerations, and constraints. CIRP Ann 65(2):737–760

Tolochko NK, Laoui T, Khlopkov YV et al (2000) Absorptance of powder materials suitable for laser sintering. Rapid Prototyp J 6(3):155–160

Zhang Y, Bernard A, Harik R et al (2017) Build orientation optimization for multi-part production in additive manufacturing. J Intell Manuf 28(6):1393–1407

Chapter 4
Electron Beam Powder Bed Fusion

Abstract Electron beam powder bed fusion (EPBF) is a process in which an electron beam is used to scan a powder bed. In order to comprehend beam-powder bed interactions and the role of various process parameters such as scan speed and beam power, it is essential to know how a beam is generated and manipulated. This chapter describes the process in detail and clarifies the roles of electric current and voltage. The process competes with selective laser melting (SLM), and it is of interest to know how this process is different. Detailed difference between EPBF and SLM is given.

Keywords Melting · Electron beam · Magnetic field · Beam-powder interaction · Laser

4.1 Process Description

Electron beam powder bed fusion (EPBF) is powder bed fusion in which an electron beam (e-beam) is used to fuse powders. It is known as electron beam melting (EBM), selective electron beam melting etc. It is commercialized by a Swedish company named Arcam (Arcam 2018). Research is done by using either commercial machines or other machines built in-house at several universities.

A schematic diagram of the process is given in Fig. 4.1 that shows the following components to realize the process: (1) beam generation, (2) beam manipulation, (3) vacuum chamber and (4) powder bed processing (Kahnert et al. 2007). The roles of various components are given below.

S. Kumar, *Additive Manufacturing Processes*,
https://doi.org/10.1007/978-3-030-45089-2_4

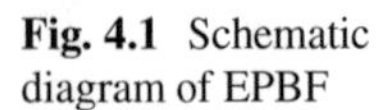

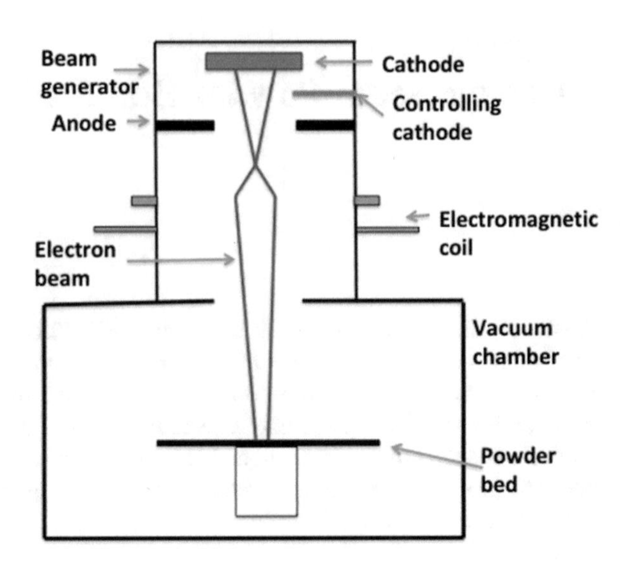

Fig. 4.1 Schematic diagram of EPBF

4.1.1 Beam Generation

Electron beam (e-beam) is required to melt and sinter powders. It is generated by electrically heating an electrode (called cathode) made from materials having high melting point and low work function. Due to high melting point, cathode will not melt and degrade during electron generation, while due to low work function, electron will be detached from cathode material at a low applied voltage. Work function is a physics term which states how much voltage needs to be applied on the material to detach an electron from its surface. Lower the work function, better the material for e-beam generation. Tungsten (W) and Lanthanum Hexaboride (LaB_6) are used as cathode materials, each of which has melting point and work function as 3422°C, 4.5 eV and 2210 °C, 2.5 eV, respectively. W is used because of its high melting point, while LaB_6 is used due its lower work function and higher life time as cathode (Edinger 2018). Besides generating e-beam by heating a cathode, which is called thermionic emission, e-beam can also be generated without heating the cathode but by creating plasma between the cathode and the anode (Bakeev et al. 2018); the beam then can be used for processing in AM (Lee et al. 2017).

Generated electron, being a negatively charged (charge = -1.6×10^{-19} Coulomb, mass = 9.1×10^{-31} Kg) ion, will move towards positively charged electrode (anode); movement of electrons towards anode will constitute an e-beam. Higher the number of electrons, higher the current of the e-beam. In order to have higher current, a large number of electrons need to be generated. In order to have lower current, either a smaller number of electrons need to be generated or some electrons need to be stopped from reaching the anode. Placing another cathode just before the anode will serve the purpose as the cathode will displace the electron and will decrease the current. Anode is made perforated so that electron will not stop at the anode but will pass through it to be available for being manipulated and for processing the powder bed.

The speed of the electron depends on the voltage applied between the cathode and the anode. Higher the voltage, higher the velocity of the electron as per relation (4.1):

$$v = c \times \sqrt{1 - \frac{1}{\left(1 + \frac{e \times V}{m \times c^2}\right)^2}} \tag{4.1}$$

where v is the velocity of the electron, c is the speed of light, e is the charge of the electron, V is the applied voltage and m is the mass of the electron (Sigl et al. 2006). For an applied voltage of the order of kV, the speed of the electron is of the order of the speed of light, bringing relativistic effect in mass or momentum of the electron, which will increase by $\frac{1}{\sqrt{1 - \frac{v^2}{c^2}}}$.

Power of an e-beam is given by relation (4.2):

$$P = V \times I \tag{4.2}$$

where V is the applied voltage and I is the current of the e-beam. Power can be increased either by increasing voltage or current. With an increase in voltage, electrons will move faster and go deeper inside the powder bed while with an increase in current, more electrons will reach the powder bed. Consequently, for the same value of power, there will be a number of different physical effects on the powder bed depending upon the value of V and I. This is the reason why V and I rather than power are fundamental variables in EPBF. For increasing the power, increasing V rather than I is a better option because with an increase in I, number of electrons will increase giving rise to an increased charging of the powder bed, causing a disturbance on the bed. Increasing V will cause an increase in the momentum of the electron but due to huge difference in mass of an electron and a powder and small e-beam-powder interaction time, the impinging electron will not be able to displace a powder causing no disturbance on the powder bed. Though, high velocity of the electron will cause an increase in the mass of the electron due to relativity, the increase is still miniscule to make a difference in an EPBF setup.

4.1.2 Beam Manipulation

An e-beam generated from a cathode and passing through a perforated anode will strike a limited area of powder bed-area equal to the beam spot size of the order of millimetre or less. It will not help making big parts (bigger than the beam spot size) unless e-beam generator moves relative to the powder bed or the bed moves relative to the beam generator. Though, making parts by these types of relative movements could be helpful to a limited extent (due to the need of maintaining a vacuum environment) in a research environment for investigating beam-material interaction or effect of various beam parameters on various materials, it has limited utility in a commercial application because of the lack of accuracy and speed.

Another option is to displace the e-beam from its original path so that it will cover wider area of the powder bed than the area equivalent to the spot size and will be able to make bigger parts. As e-beam consists of charged particles which will be displaced by an applied electric field; moving charged particle (i.e. an e-beam) creates a magnetic field and will be displaced by an applied magnetic field – showing an electromagnetic field is a solution.

An electron of charge e moving with velocity v will feel a force F, named Lorentz Force in an electromagnetic field as per following relation (4.3):

$$F = e \times (E + v \times B) \tag{4.3}$$

where E denotes an electric field and B denotes a magnetic field. F acts perpendicular to the plane made by v and B. Figure 4.2a shows x, y and z directions in a cartesian coordinate.

In order to comprehend the role of magnetic field on a moving electron, electric field is considered negligible which changes Eqs. (4.3) and (4.4):

$$F = e \times (v \times B) \tag{4.4}$$

Expanding Eq. (4.4) in terms of its components in x, y and z directions, it can be shown by Eqs. (4.5, 4.6 and 4.7):

$$F_x = e(v_y B_z - v_z B_y) \tag{4.5}$$

$$F_y = e(v_z B_x - v_x B_z) \tag{4.6}$$

$$F_z = e(v_x B_y - v_y B_x) \tag{4.7}$$

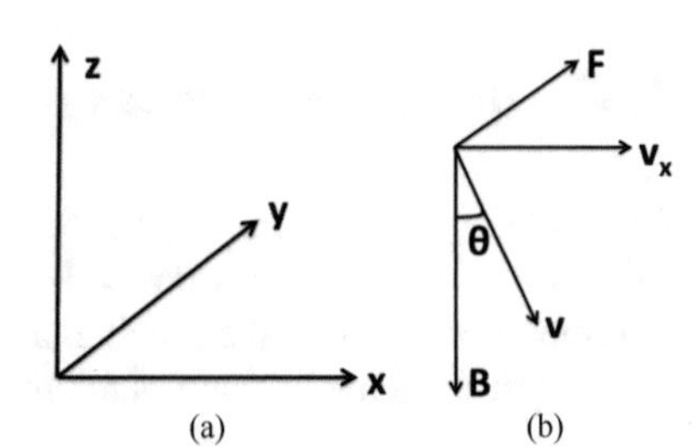

Fig. 4.2 (**a**) Cartesian coordinate, (**b**) Lorentz force F on an electron moving with velocity v in a magnetic field B

In order to find resulting force on an electron moving in z-x plane with velocity v at an angle θ with B which is acting in direction z as shown in Fig. 4.2b, following information are used in Eqs. 4.5, 4.6 and 4.7:

$B_y = 0$, $B_x = 0$ since B is in z direction, $v_y = 0$ since v is along z-x plane, it is found:

$$F_y = -ev_x B_z \quad \text{while} \quad F_x = 0, \quad F_z = 0$$

$$\text{Since, } v_x = vSin\theta, \quad B = -B_z \quad F = F_y = -ev_x B_z = evBSin\theta \tag{4.8}$$

Equation (4.8) states the direction and magnitude of the Lorentz force acting on the moving electron; the force is in y direction and acting on z-x plane as shown in Fig. 4.2b.

The consequence of this force on the moving electron can be known by observing the direction of the force. Since the magnetic field acts in circular direction on a moving charged particle (electron), the magnetic field is acting in clockwise direction on the electron to cause resultant force in y direction; it is similar to a rotation of a screw which if rotates in clockwise direction from v to B through smaller angle in z-x plane (Fig. 4.2b) will cause an advancement in positive y direction. Vice versa, if there is a force in y direction there has been rotation of electron clockwise. Observing the direction of F in Fig. 4.2b, it can be known whether electron direction is clockwise or counter-clockwise; clockwise implies displacement of the electron towards magnetic field and counter-clockwise means the displacement to be away from the magnetic field.

In the present case, as per Fig. 4.2b, the displacement of electron is towards magnetic field. How far the electron will displace will depend on the magnitude of v and B; if B is large or there is a strong magnetic field, force will be large enough on the electron to align it with the magnetic field; with an alignment, angle θ will be zero and as per Eq. (4.8), force will be zero causing the electron to move indefinitely parallel to the magnetic field. This concept is utilized in focussing an e-beam in EPBF.

Using $\theta = 0$ in Eq. (4.8), $F = 0$, it gives another inference that when magnetic field is applied parallel to the e-beam, it has no effect on the electron while for any other value of θ, the beam is deflected; it implies that a magnetic field cannot be used to accelerate an electron along optical axis; acceleration can only be accomplished by an applied voltage.

In case, magnetic field B is reversed so that $B = B_z$, it will change the equation to $F_y = -ev_xB$, which is acting in negative y direction normal to z-x plane implying a counter-clockwise motion of vector v; it will make the angle θ bigger. This shows that by reversing the magnetic field direction, electrons will be displaced away from the magnetic field. This will help e-beam to move away from its original path and scan wider length. By fast fluctuating magnetic field direction from positive to negative, e-beam direction will change fast while changing the strength of the magnetic field, e-beam velocity on the powder bed (scan speed, Fig. 4.3) can be changed. This shows that there are two ways by which scan speed on the powder bed can be altered – by changing either applied voltage or magnetic field. For scanning along more axes, more electromagnets are required as sources of magnetic fields. This concept is utilized in scanning e-beam in EPBF.

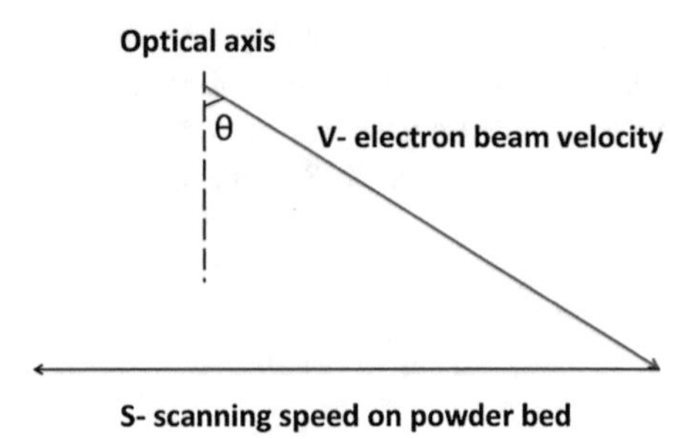

Fig. 4.3 An e-beam with velocity v is scanning powder bed with scan speed s

Figure 4.3 shows scanning of a powder bed along a line using e-beam moving with velocity v. Scan speed s does not depend upon the e-beam velocity but depends upon the speed with which the beam is deflected towards optical axis, which depends upon the magnitude of magnetic field. If a high magnetic field is applied, e-beam travels fast on the powder bed imparting less number of electrons on the bed. If the field is low, scanning speed will be slow imparting high number of electrons on the bed, resulting in high effective current on the bed, implying higher energy density applied on the bed.

For constant applied voltage (60 kV in case of Arcam EBM systems), changing applied current is a means to control number of electrons on the bed; if scan speed is set higher resulting in lower number of electrons, applied current can be increased to compensate a decrease in electron numbers resulting in no change in current density with an increase in scan speed. Thus, adjusting scan speed and applied current can be used to maintain desired energy balance on the bed. Finding right energy density by making an adjustment towards higher scan speed will result in higher production rate, while for obtaining higher accuracy of parts such adjustment can be overlooked.

4.1.2.1 Focussing of Beam

In EPBF, focussing of an e-beam is done by applying a magnetic field parallel to the optical axis along which the beam is expected to travel. The direction of the magnetic field, optical axis and velocity of the electron is shown in Fig. 4.4a. E-beam after being accelerated by an applied voltage does not exactly move in a straight line along the optical axis but moves in a curved path due to the presence of other electromagnetic fields in EPBF (as shown in Fig. 4.4a. If an e-beam entering the magnetic field makes an angle with the optical axis, it will be deflected by the field (as shown in Fig. 4.2b) and will become parallel with the optical axis as shown in Fig. 4.4b.

Applying a magnetic field parallel to the optical axis helps achieve focussing of the beam but it can also be accomplished by applying the field perpendicular to the axis as shown in Fig. 4.5. If the beam is deflected from the optical axis, then it can be brought back to the same path by applying B perpendicular to it as shown in Fig. 4.5a, in which a beam is making an angle θ with B; the magnitude of the field should be set to a particular value so that the resulting force will deflect it by an angle $\theta - 90^{\circ}$ and align it with the axis, causing focussing of the beam. In order to

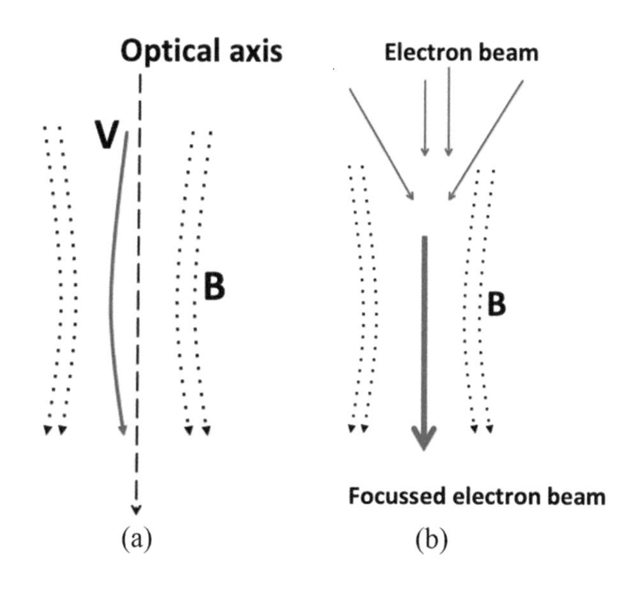

Fig. 4.4 An electron moving with velocity v in a magnetic field B: (**a**) approximate beam path (Azhirnian and Svensson 2017), (**b**) schematic diagram of focussing of electrons entering the field at different angles

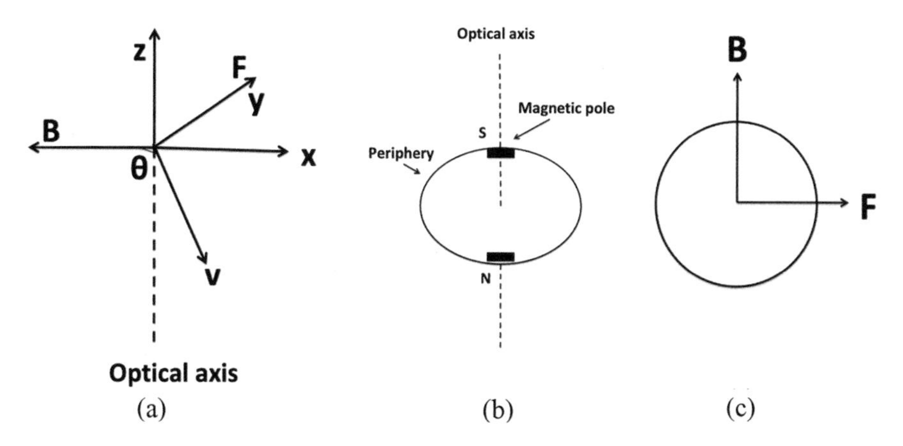

Fig. 4.5 Focussing of an e-beam by deflection: (**a**) direction of B, F and v in cartesian coordinate, (**b**) position of electromagnet on the periphery of e-beam column, (**c**) direction of B and F

execute it, two magnets are attached on the periphery of an e-beam column – north pole of one magnet while south pole of another magnet is facing the periphery as shown in Fig. 4.5b, which generates a magnetic field; the direction of the field and force are perpendicular to each other as shown in Fig. 4.5c. These magnetic poles will displace the beam along x direction; in order to displace it along y direction, two more magnetic poles need to be set on the periphery normal to the fitted poles. These four magnetic poles will force align any e-beam making any angles with the optical axis, causing focussing of the beam.

4.1.2.2 Beam Shape Control

Beam during scanning may not retain its original shape and deform to various shapes. Figure 4.6a shows a beam of original spherical shape which deforms resulting in elongation along AB while compressing along CD as shown in Fig. 4.6b. The beam can be brought to its original shape if compressive force acts along AB, and pulling force acts along CD as shown in Fig. 4.6c. Electrons at point D of the beam need to be deflected outwards while electrons at point A need to be deflected inwards to bring them back to their original position. As per Fig. 4.5a, the former deflection can be achieved by applying B along x axis, orthogonal to the optical axis while the latter inward deflection can be achieved by applying another magnetic field in negative direction along y axis – this requires an arrangement of four magnetic poles.

This can be achieved by a quadrupole magnet, as shown in Fig. 4.6d, which applies the force exactly in the same direction as required (Azhirnian and Svensson 2017); the figure shows, four poles are placed alternatively and magnetic lines are travelling from N to S creating two opposite forces in x and y axis. For higher degree of deformation, higher number of poles of equal number of similar poles are required.

4.1.3 *Vacuum Chamber*

A vacuum chamber is an essential part of EPBF, in which e-beams travel and interact with the powder bed; in absence of vacuum, electrons will interact with air molecules and ionize them and thus losing energy, speed and direction during travelling, causing difficulty in processing the bed. In Arcam systems, a vacuum of 10^{-5} mbar is maintained which is sufficient to allow unhindered movement of e-beams. Another advantage of vacuum is that it enables reactive metals to be processed without being reacted by reactive gases such as oxygen or nitrogen (Korner 2016). Vacuum also gives disadvantages: it decreases the melting point of metals and increases the rate of evaporation of metals; for an alloy made up of two different metals having different melting points, the rate of evaporation of lower melting point will be higher resulting in its higher loss; it will change the final composition

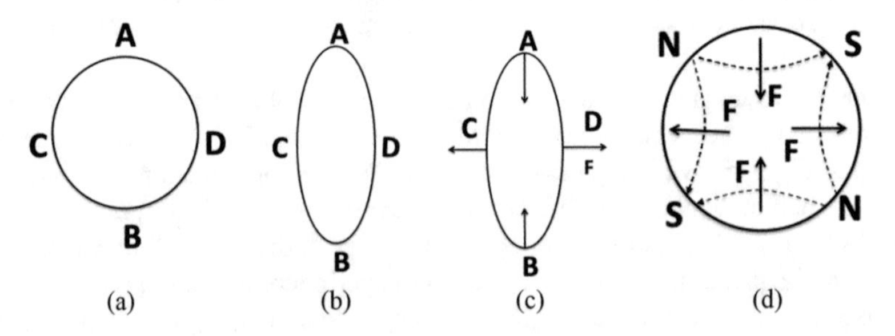

Fig. 4.6 (**a**) Original beam shape, (**b**) deformed beam shape, (**c**) direction in which force needs to be applied for regaining the shape, (**d**) quadrupole magnet as a device to apply required force

of the alloy. This has been noticed in case of Ti6Al4V alloy, where higher loss of Al changes the final composition of titanium alloy. Another disadvantage of vacuum is that it takes time to create vacuum, which decreases the production rate. Irrespective of its advantage or disadvantage, vacuum is a necessity in EPBF.

Interaction of e-beam with powders leads to transfer of electrons to powders, causing charging of powders. Accumulation of negative charges on powders has potential to destabilize powders' positions and bring disruption in layer formation. This is due to coulombic repulsion between two same charges; when charges are high, the resulting repulsion is higher than the inertia of powders leading to the displacement of powders. Presence of helium gas (10^{-3} mbar partial pressure in Arcam systems) in the vacuum chamber mitigates this effect (Korner 2016). Helium has low atomic number, and, at a pressure of 10^{-3} mbar, the number of helium atoms is not large enough to disturb e-beams. But these numbers are big enough to take substantial amount of charge away from charged powders by being in vicinity of powders, causing mitigation of disruptions of the bed.

4.1.4 Powder Bed Processing

Since powder bed gets disrupted during its interaction with e-beams due to charging of powders, processing of powder beds in EPBF requires special attention. Application of heat is the best way to mitigate the effect of charging, which is described as follows.

After layer formation, if powders get attached to the base plate or the previously deposited layer, then incoming electrons will not be able to accumulate on the powder bed and will dissipate away. Attaching the powder can be done either by sintering or melting it; melting or partial melting of powders cannot be an option as all powders which are not melted by e-beams in EPBF need to be detached and removed from the final part. This leaves another option, that is sintering to be exercised for making connection among powders and with the surroundings. Sintering implies that the powder will join another powder or base plate or previously deposited layer by forming a neck through diffusion of atoms. Formation of neck increases electrical conductivity of the powder layer and enhances contacts between substrate (or underlying layer) and powders, which helps dissipation of electric charges to the ground – resulting in removal of cause (charges) responsible for displacement of powders. A minimal size of neck obtained after some second of sintering is sufficient for charge transfer and stabilization of the powder bed. Since the role of the neck is to act as a conduit to transfer charge rather than to act as a fixture to hold the powder, larger powder (Cordero et al. 2017) because of its higher inertia rather than a smaller or fine powder has a higher chance not to be displaced during charging; this makes larger powder rather than smaller powder more suitable for EPBF.

Sintering is accomplished by application of heat using heaters and e-beams. Heater attached under substrate is used to increase the temperature of the substrate over which layer is deposited. When powder layer comes in contact with the substrate, they become sintered. Lower temperature of the heater increases the sintering time and consequently decreases the production rate. Higher temperature decreases the sintering time and consequently increases the production rate; however, higher temperature of the vacuum chamber increases the sintering of unmelted powders of previously deposited layers – resulting in difficulty in removal of the powders from the part and compromising fine features. This can be avoided by use of low heater temperature and increasing the temperature of the deposited layer with the help of e-beams. This will ensure a no decrease in production rate as well as non-occurrence of over-sintering. For increasing the local temperature of powder layer, high power e-beams with high speed are scanned over whole layer several times. This will cause necessary sintering. An e-beam moving at high speed does not impart enough electrons at a point to cause disturbance on the layer. E-beam for sintering purpose can be used at several steps – on the substrate (for replacing the heater), on the substrate (for helping the heater) before layer formation on the deposited and processed layer, after layer formation, after processing the powder layer; this depends on how operator wants to reach optimized production rate and sintering. All options will lead to creation of a hot processing chamber. This type of hot chamber resulted due to sintering and melting has several advantages such as decrease in thermal stress, cracks, porosity and several disadvantages such as increase in ductility, grain size, surface roughness etc. Irrespective of advantages and disadvantages it affects, a hot chamber in EPBF is a basic necessity and cannot be avoided.

4.2 Difference from Selective Laser Melting

4.2.1 Powder

In EPBF, powder needs to be electrically conductive to facilitate dissipation of charge from it to the surroundings which is in contact with it. This limits the powder that can be processed in EPBF; mostly metal and alloys are used. Powders which form oxide or nitride layers on their surface due to contamination with gases diminish their suitability for EPBF, because these layers have lower electrical conductivity.

In selective laser melting (SLM), all materials, metals, ceramics, polymers, composites, are in principle processable. The number of processable SLM materials are low; it is not because there is a yardstick (such as electrical conductivity in EPBF) available lack of which make the materials unprocessable but because less number of materials are investigated.

4.2.2 Beam

In case of conversion of electrical energy into high energy beam (electron or laser), also known as wall-plug efficiency, the rate of conversion for e-beam is about 95%, while for laser beams, used in SLM, it is from 5% to 20% for CO_2 laser, up to 20% for Nd: YAG laser, from 30% to 40% for fibre laser, up to 70% for diode laser (Hecht 2018; Li 2000) showing higher efficiency of e-beam than laser beams.

4.2.3 Beam-Powder Interaction

E-beams are made up of electrons which when irradiated on metal powders get deflected by free electrons surrounding metal atoms. If the power of e-beam will be higher, it will not be deflected by free electrons but will move deeper inside until it is decelerated by lattice of atoms and stops. Lattice vibrates and generates heat; thus, kinetic energy of moving electron is transferred into heat energy, responsible for melting. If the size of atom is big (higher atomic number), there is a less chance for e-beam to escape, and thus bigger atom size increases the efficiency of EPBF. For an applied voltage of the order of kV, the size of e-beam impingement in powder is of the order of microns.

Laser beam is made up of photons which are neutral particles; when it irradiates metal powders, it interacts with free electrons surrounding atoms; for a high number of free electrons, photon has a chance to be reflected. For a laser beam of high intensity or metal having fewer number of free electrons, incoming photons will interact with bound electrons; the electrons will re-radiate or interact with lattice atoms causing vibration. Photons deflected by bound electrons have less chance to escape (than the photons deflected by free electrons) and will reflect internally until it loses energy by interacting with lattice. Thus, incoming laser energy will be transferred to metal powders as heat energy. When incoming photon reaches up to bound electrons without being deflected by free electrons, it is able to transfer its energy; this distance inside the powder (penetration depth) is about few nm (Fig. 4.7). This shows that laser beam-material interaction is a surface phenomenon while e-beam-material interaction (penetration depth- few micron) is a relative bulk phenomenon.

While a laser beam is prone to be reflected from the metal powder, there is no such disadvantage with e-beam – making e-beam more efficient in terms of transfer of energy to metal powders as well as in terms of generating it from electrical energy. The comparative inefficiency of laser beam is somewhat compensated in powder bed processing when reflected beam has a chance to be directed towards another powder and be absorbed and utilized; this depends upon the surface roughness of powders, gap between powders and angle of incidence.

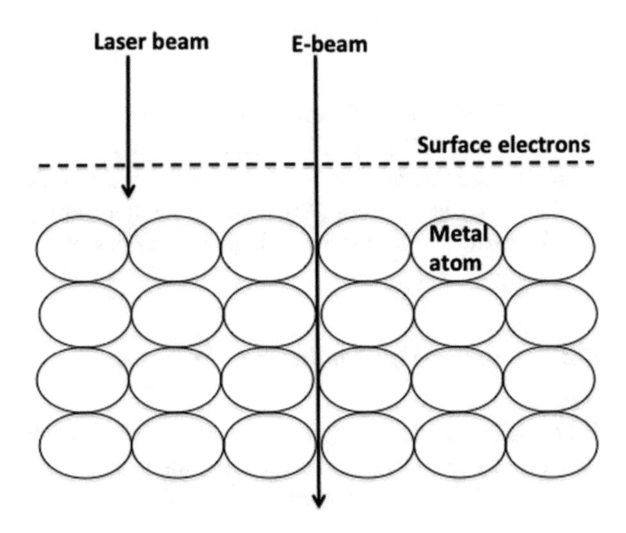

Fig. 4.7 Relative penetration depth of laser and electron beams in a metal lattice

Metals such as copper, aluminium, silver having high reflectivity are not suitable to be processed in SLM, while these are suitable for EPBF. These reflective metals if developed an oxide layer such as alumina layer on aluminium will no longer be suitable for EPBF because of charge accumulation while these are suitable for SLM because of their decreased reflectivity. This is a classic example of demonstrating what is good in one process may not be good in other process and vice versa – implying both processes will well-complement each other.

4.2.4 Parameter

Laser beam moves with a constant speed, that is speed of light, while e-beam has varied speed depending upon the applied voltage. Applied energy density in EPBF can be changed by changing the speed of e-beam while there is no such option in SLM as the speed of light cannot be varied.

Laser power is an independent quantity; for a given setting of laser power in SLM, it does not change with a change in other experimental parameters such as scan speed and scan strategy. In EPBF, power may change with a change in scan speed or scan strategy. This is the reason why keeping power constant and changing other experimental parameters may give useful information in SLM, the same may give misleading information in EPBF, if all details are not taken into account.

SLM can be carried out using a laser beam of different types of wavelengths (such as YAG laser, fibre laser); different wavelengths have different absorptance for different materials. There is no such provision in EPBF that it can be carried out using an e-beam of a particular wavelength. It may be argued that an e-beam accelerated at an applied voltage has a particular wavelength (de Broglie wavelength), but during EPBF, the speed of beam varies bringing variation in wavelength; it implies that there is no EPBF with a fixed wavelength though there is SLM with a fixed wavelength.

4.2.5 Implication of Vacuum

Vacuum increases the evaporation rate of materials and changes the composition of processed materials, which implies that SLM (without vacuum, with inert gas) will cause lesser composition change than EPBF. Vacuum has advantage there is no gas in the chamber and therefore no gas will be absorbed by liquid metal pool, absorption of gas is a source of porosity.

SLM has an edge over EPBF to process the material by increasing the atmospheric pressure in the chamber; high pressures will decrease the porosity, surface roughness (Bidare et al. 2018) and sublimation. Besides in SLM, gas flow inside the processing chamber can be managed to change surface roughness and quality (Montgomery et al. 2018); flow parallel to the powder bed is used to remove particulates from the chamber and is used as a parameter to vary the properties (Ferrar et al. 2012).

4.2.6 Scanning

EPBF works using electromagnetic coils based scanning system which enables scan speed of the order of 1000 m/s while SLM works on mirror based scanning system which provides scan speed of the order of 10 m/s. EPBF, due to its higher scan speed, furnishes distinct advantages over SLM such as potential for higher production rate, mitigation of thermal stress by use of simultaneous parallel beams, utilization of beam as a source of an additional heater etc.

In EPBF, high scan speed combined with high beam power enables transfer of high energy density at a faster rate leading to faster melting of whole layer; this fulfils one of the tasks for higher production rate. SLM has lower scan speed and does not match EPBF in this aspect. However, there is a limit by which scan speed in SLM or EPBF can be increased. At higher scan speed, there will be small interaction time for a beam with powder (of the order of nanosecond for a beam size of 100 micron moving with 1000 m/s); this will cause full or partial ablation (Steen and Majumder 2010) generating evaporated materials interfering with vacuum and beam (particularly in EPBF), pressing the molten material by recoil pressure, non-melting of powders and finally non-realization of the process. Scan speed more than a certain speed cannot be employed; consequently, a beam power more than a certain power cannot be employed in a scan. This is the reason why the laser power of several kW is available for decades but there is no single SLM system available matching that power. Though, higher beam power can be utilized by distributing the power among several beams with a restriction that the power in a beam should not exceed the power limit. If the processing chamber is hot, as in case of EPBF, still lower power (or applied energy density) is required to raise the temperature of the layer and bring it to the level of melting point.

Employing a number of beams simultaneously to scan a layer is another method to increase the production rate. For parallel beams scanning a layer, if their heat-affected zones are either touching or overlapping each other, the development of thermal stress across the layer will be minimal; in this way, parallel scanning not only increases the production rate but also minimizes the thermal stress.

References

Arcam AB (2018). www.arcam.com
Azhirnian A, Svensson D (2017) Modeling and analysis of aberrations in electron beam melting (EBM) systems. Master thesis, Chalmers University of Technology
Bakeev IY, Klimov AS, Oks EM, Zenin AA (2018) Generation of high density electron beams by a forevacuum- pressure plasma cathode electron source. Plasma Sources Sci Technol 27:075002
Bidare P, Bitharas I, Ward RM et al (2018) Laser powder fusion in high-pressure atmospheres. Int J Adv Manuf Technol 99:543–555
Cordero ZC, Meyer HM III, Nandwana P, Dehoff RR (2017) Powder bed charging during electron-beam additive manufacturing. Acta Mater 124:437–445
Edinger R (2018) Laser heated electron beam gun optimization to improve additive manufacturing. In: Proceedings of the solid freeform fabrication symposium, Austin, TX, USA, pp 2297–2304
Ferrar B, Mullen L, Jones E et al (2012) Gas flow effects on selective laser melting (SLM) manufacturing performance. J Mater Process Technol 212:355–364
Hecht J (2018) Understanding lasers: an entry-level guide. Wiley, Hoboken
Kahnert M, Lutzmann S, Zaeh MF (2007) Layer formations in electron beam sintering. In: Proceedings of the solid freeform fabrication symposium, Austin, TX, USA, pp 88–99
Korner C (2016) Additive manufacturing of metallic components by selective electron beam melting – a review. Int Mater Rev 61(5):361–377
Lee HJ, Ahn DG, Song JG et al (2017) Fabrication of beads using a plasma electron beam and satellite 21 powders for additive manufacturing. Int J Precis Eng Manuf Green Tech 4:453
Li L (2000) The advances and characteristics of high-power diode laser materials processing. Opt Lasers Eng 34(4–6):231–253
Montgomery C, Farnin C, Mellos G et al (2018) Effect of shield gas on surface finish of laser powder bed produced parts. In: Proceedings of the solid freeform fabrication symposium, Austin, TX, USA, pp 438–444
Sigl M, Lutzmann S, Zaeh MF (2006) Transient physical effects in electron beam sintering. In: Proceedings of the solid freeform fabrication symposium, Austin, TX, USA, pp 464–477
Steen WM, Majumder J (2010) Laser material processing. Springer, London

Chapter 5
Other Powder Bed Processes

Abstract Beam based powder bed fusion makes complex parts, but the process is slow, energy-inefficient and is not cost-effective to make low-value parts. Powder bed processes such as high speed sintering, selective heat sintering, binder jet three-dimensional printing and other emerging processes (micro heater array powder sintering, localized microwave heating based additive manufacturing, multi jet fusion) are more energy-efficient and cost-effective – the present chapter describes these processes. There are various types of scanning such as pointwise scanning, linewise scanning and areawise scanning which impact fabrication rate and resolution. Difference between these types of scanning is explained.

Keywords High speed · Binder jetting · Microwave · Heater · Areawise scanning · Fusion · Inhibiter

5.1 Introduction

Beam based powder bed fusion makes complex parts but it is expensive because beam is expensive. If the beam will be removed, then the process will not be expensive. This brings a question what a beam does so that an alternative arrangement could be done for replacing the beam. Beam heats a powder bed. If heating of the powder bed is required, then a resistance heater can do the job (Baumers et al. 2015). But, the heat coming from a heater will go everywhere and is not as directed as a beam is. Then, the heater can be placed close to the powder bed so that going of the heat to everywhere could be minimized. But, the beam can be deflected and whole powder bed can be raster scanned; and in case of a heater, either the heater needs to be moved or the powder bed needs to be moved for the whole powder bed to be heated – which does not give as convenience as a beam gives. Then, a thousand tiny heaters can be arranged all over close to the powder bed so that by switching these heaters on and off, whole powder bed can be selectively heated – this could be a new process which does away with the beam and does away with the expense as well (Holt et al. 2018). This brings a basic question – why heating the powder bed

S. Kumar, *Additive Manufacturing Processes*,
https://doi.org/10.1007/978-3-030-45089-2_5

is required or what is the purpose of heating the powder bed. Heating is done to join powders. If joining the powders is the only purpose, then heater might not be indispensable, spraying the powder with glue (or ink or binder) instead can serve the purpose (Enneti et al. 2018; Kernan et al. 2007). Thus, there could be another process where not only the beam but also the heater is done away with in favor of an ink jetter or a binder jetter. This chapter describes such types of powder bed processes.

5.2 Non-beam Based Powder Bed Fusion

Beam based powder bed fusion (PBF) is widely researched because beam provides high energy density and resolution; high energy density allows to process low melting point materials (such as polymers) as well as high melting point materials (such as metals and ceramics) while resolution allows to create fine features and to introduce details in a part. But, beam based PBF has some demerits – it is slow in processing, it is expensive, it requires much electrical energy to generate beams – these motivate to explore other processes not based on high energy beams, which will then be free from such demerits. Non-beam heat sources are infrared lamps, concentrated microwave energy and resistance based heaters. Microwave energy controlled by localized microwave heating applicator is not considered as a beam because it cannot be manipulated or deflected as a beam, besides the applicator needs to be placed near the powder bed surface; the applicator like electron beam sources or laser beam sources cannot be placed away from the powder bed surface (Jerby et al. 2015). Some non-beam based PBF are given below.

5.2.1 Heater Based Sintering

Beam has quality that it can directed to a point. If a heater is used in place of a beam source, then heat radiation coming from the heater will start to diverge. Thus, if a heater is placed above a powder bed, then the radiation coming from the heater will not converge at a point below the heater on the powder bed. If the separation between a heater and a powder bed will increase, the effect of divergence will be more clear, the heat energy density will further decrease. If the aim is to melt powder with the help of a heater, then the heater must be kept in close proximity to the bed, which may damage the heater itself if not protected. It works well with low melting point materials such as polymer for which lesser heat-induced risk is involved. Beam has no such disadvantage but is expensive. Utilizing heater instead is a low cost effort to try to achieve the same. Beam has quality to be deflected and thus it can cover the whole area of a layer without having a need to move its source. Heater has no such quality; therefore, it needs either to move to cover whole area or to be present everywhere so that whole area is already under its reach. If heater needs to move, then it reminds the movement of an ink jet printhead which also covered the whole area without having the quality of deflection. An ink jet printhead consists of a number of ink jets, if the printhead instead will consist of a number of small heaters then it will serve the purpose. This is the concept behind selective heat sintering (SHS) (Baumers et al. 2015) (Fig. 5.1). If heater needs to be present everywhere, then a number of small heaters need to be fitted in a cover or mask covering the whole area, this reminds the presence of a number of small mirrors in digital light processing. Selecting some of the heaters switched on while others switched off will help print selectively. This is the concept behind micro heater array powder sintering (MAPS) (Holt et al. 2018).

5.2.2 *Localized Microwave Heating Based AM*

Microwave is an electromagnetic radiation having wavelength from 1 mm to 1 m while laser used in AM is of the order of μm (Pinkerton 2016). Microwave is used in sintering (Leonelli et al. 2008) as well as in post-processing of AM parts (Salehi et al. 2019) because it saves time and energy for processing. While other radiations (laser) are absorbed on the surface of a powder giving rise to surface absorption, microwave gives rise to volumetric absorption. Due to non-linear effect of absorption, temperature of powders can increase drastically after some time giving rise to high temperature at low input microwave energy. This high rise in temperature, which is called thermal-runaway effect, makes microwave not only a potential thermal source for high temperature processing but also a source having unpredictable behaviour at different conditions which requires to be controlled. In microwave-assisted selective laser melting, it has been studied as a complementary heat source to laser beam so that in the presence of microwave, low laser power scanning will take advantage of thermal-runaway effect and will do the job of high laser power. This will help higher melting point ceramics to be processed which otherwise furnished cracked parts (Buls et al. 2018). Application of electric field across powder bed is also found to be useful to decrease sintering time and temperature of ceramics (Hagen et al. 2018).

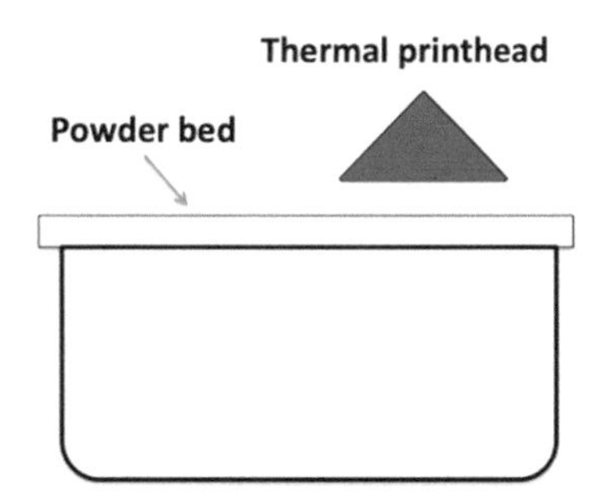

Fig. 5.1 Schematic diagram of selective heat sintering (SHS)

Application of microwave in PBF as a primary heat source requires it to be concentrated at a small area so that its effect could be realized at the desired zone of a processing area without damaging other parts of a PBF system. Application of a microwave applicator or microwave localized heating (LMH) applicator has demonstrated the localized melting of metal powders. Using the applicator, microwave of wavelength ~12 cm (2.45 GHz) can be concentrated to a diameter of ~ 1 mm; this is in sharp contrast to laser where spot size lower than the wavelength is limited by diffraction (Jerby et al. 2015).

5.3 High Speed Sintering

High speed sintering (HSS) is non-beam based PBF in which each layer is scanned twice- first, a shape is defined on a layer by an ink jet printhead (Fig. 5.2a); second, the whole layer is irradiated by a thermal lamp to consolidate that shape (Fig. 5.2b) (Brown et al. 2018; Thomas et al. 2006).

The ink jetted by a printhead is a radiation-absorbing material, and thus the area on the layer where ink is jetted absorbs more thermal radiation than other adjacent areas. The role of the printhead is thus to mark areas on a layer which will constitute a part. The role of the thermal lamp is not to mark any area and thus does not need to be aware of the details of a CAD file of a part. The lamp is thus free from a tool path; it just covers the whole layer from one side to another side. The movement of a lamp is thus relatively fast. In a few seconds, the lamp can scan whole layer starting from one end of a layer; this speed of the lamp has given the name of the process 'high speed sintering'.

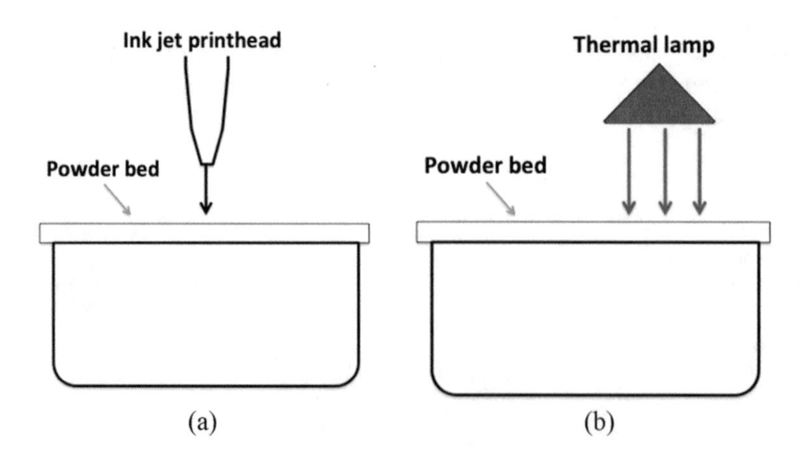

Fig. 5.2 Schematic diagram of high speed sintering (HSS): (**a**) first step: ink jetting, (**b**) second step: irradiation by thermal lamp

The speed of the process is summation of two factors: (1) time taken by ink jet printhead for jetting on a layer and (2) time taken by lamp for scanning the layer; this speed is faster than that of a comparative process such as selective laser sintering (SLS). Scanning by laser in SLS is slow while jetting by a printhead is fast, therefore printing speed in HSS is faster than SLS. Though, for processing a layer in SLS, only one step (laser scanning) is required while for the same, two steps (jetting plus lamp scanning) are required in HSS. Time taken by lamp scanning is insignificant in comparison to the time taken by jetting, and therefore the speed of the process is mostly influenced by the speed of jetting. However, if the size of a part is small, then HSS will no longer be having advantage over SLS in terms of speed, or HSS will no longer be faster than SLS. It is because there will not be a significant difference between the time taken by laser scanning and the time taken by ink jetting while the time taken by lamp scanning will not be insignificant. On the contrary, if the size of a part is big or if the combined size of many small parts is big so that processing to complete a layer will take longer time, then HSS will be having advantage over SLS in terms of speed, which will justify 'speed' of high speed sintering (HSS).

In a variant of HSS named multi jet fusion (MJF), two types of inks are used – one for enhancing absorption of the radiation (same as that used in HSS), while the second ink is an inhibiter type ink used at boundaries of a scanned pattern. The role of the second ink is to decrease the diffusion of heat outside the boundary, if heat will be diffused outside then it will cause some powder particles to join outside the boundary resulting in a rough side surface. Using the second ink will make side surface smooth and increase the accuracy and definition of a scanned pattern (Sillani et al. 2019). Use of two types of ink is same as two types of parameters – one for scanning at the boundary and the other at non-boundary area in beam based powder bed fusion (Tian et al. 2017).

The radiation-absorbing material is carbon black. The heat absorption by carbon black must be enough to join nearby powders; though, this absorbed heat is not high but sufficient enough to join low melting point materials such as polyamide and elastomers; this heat will not be enough to join high melting point materials such as iron or titanium. The process, thus, has limited applications. The use of carbon especially in the form of graphite is not new to enhance absorption of laser (Ho et al. 2002; Wagner et al. 2005). Presence of carbon black decreases the mechanical properties of polyamide in another AM process (SLS) (Athreya et al. 2010), or its presence does not have a positive influence on mechanical properties (Hong et al. 2019). Carbon in the form of a fibre increases both light absorption and mechanical properties (Goodridge et al. 2011; Chatham et al. 2019) but HSS does not use carbon fibre. This brings limitations to properties that could be aimed at to achieve with HSS. Absorption characteristics of infrared radiation coming from a laser source may not be similar to that coming from a thermal lamp, but the material which showed negative properties in SLS platform cannot be expected to change drastically and furnish positive properties in HSS platform.

5.3.1 Energy-Efficiency of High Speed Sintering

The process does not need a laser and is therefore free from expensive investment. Laser is itself not an efficient energy-conversion device, approximately only from 5 to 70% of electrical energy is converted into laser energy (Li 2000; Hecht 2018). Most energy-efficient laser (of more than 40% efficiency) is diode laser, but it has no widespread use in AM (Pinkerton 2016; Arredondo et al. 2017) due to its poor beam quality (Santos et al. 2006). Since HSS uses a thermal lamp which is almost 98% efficient (~2% is lost in non-thermal light energy), HSS is more energy-efficient than a laser powder bed fusion (LPBF) as far as selection of heat source is concerned. In HSS, all powders of the layer get processed, majority of these powders do not end up in a part. In SLS, only those powders of the layer that will end up in a part get processed. Thus in HSS, majority of powders which are not getting ended up in a part are still getting processed, still getting directly affected by thermal radiation, are getting exposed to degradation which may follow these thermal radiation, and are continuously losing their ability to be recycled. It is argued that print speed is high and therefore exposure time is low and therefore there will be recyclability (Sillani et al. 2019), but low exposure time is not better than no exposure time. In SLS, there is no such process-induced degradation, no such process-induced risk, no such ill-treatment to powders; there is no such emergency to speed up the process at the cost of non-selection of powders in a layer. It does not mean that powders are not at all degraded in SLS, powders are certainly degraded (Kumar and Czekanski 2018), but they are not at such risk of getting as degraded as in HSS.

Thus, SLS is more efficient than HSS in non-degradation of powders and, therefore HSS is more powder-inefficient than SLS. Powder-inefficiency of HSS will increase if HSS system becomes bigger to speed up the production process and production, because the size of a layer (powder bed) will increase and therefore more powders will be at a risk of degradation. Since production of powder is an energy-intensive process (Fredriksson 2019); HSS being powder-inefficient, cannot be energy-efficient as far as utilization of powder is concerned. In this regard, SLS is more energy-efficient. If SLS is not considered energy-efficient, then it is not because the process itself is inefficient but because the process uses an energy-inefficient device (laser).

5.4 Linewise, Pointwise, Areawise Scanning

5.4.1 Linewise Scanning

In high speed sintering (HSS) (Thomas et al. 2006), when a thermal lamp starts to scan from one end of the layer then what it does – in the first few moment, it irradiates the whole width of the layer but a fraction of the whole length of the layer – it does not seem the lamp has scanned any area – the lamp has scanned just a line on the layer. After a few more moments, the lamp covers some more length, it then seems that the lamp has covered some area. The lamp starts from scanning a line, and this line becomes wider and wider with the progress of scanning, and the line becomes an area. This is called linewise scanning. Linewise scanning is faster because it starts from a line; this scanning would not have been faster if this scanning has to create a line from scratch or if it had to create a line by adding many fragments or many points. This scanning is free from such troubles; this scanning does not need to know how to constitute a line by controlling many points. Thus, this scanning does not control many points; but, this scanning does inform that there might be some scannings which might be controlling many points, which might not be linewise scanning; there might be many scannings other than linewise scanning.

Examples of linewise scanning: stacking of laser beam emitters in a bar across powder bed and processing the powder bed without deflecting beams but by switching on/off some emitters and moving the bar (Arredondo et al. 2017; Dallarosa et al. 2016), movement of infrared lamp across powder bed in multi jet fusion (MJF) (Sillani et al. 2019). HSS is an example of double linewise scanning – the first linewise scanning takes place when an array of nozzles deposits material on powder bed, and the second linewise scanning takes place when an infrared lamp moves.

5.4.2 *Pointwise Scanning*

If a scanning starts from a point then it would not be linewise scanning, it would be called pointwise scanning. If it is pointwise scanning then it has to make many overlapping points to make a single line, and thus this pointwise scanning would not be as fast as linewise scanning. But, the pointwise scanning controls a point, it has ability to add points, it has ability to add points in a fashion it wants to add. Thus, the pointwise scanning can make a line of any size, a line in any direction, a combination of many lines of many sizes in many directions. If a linewise scanning is for fast scanning, then pointwise scanning is for fine scanning; if linewise scanning is to provide high production rate, then pointwise scanning is to provide high complexity; if linewise scanning can make both a line and an area then pointwise scanning can make all- a point, a line and an area.

Examples of pointwise scannings are: using laser beam as a point source in SLS, SLM; using electron beam as a point source in EBM (Gibson et al. 2010); a single ink jet source in BJ3DP.

5.4.3 *Areawise Scanning*

A linewise scanning starts from a line and ends up making an area, a pointwise scanning starts from a point and ends up making an area. If they do not make an area, a part will not be made. How they start is important but whether they are able to end up in an area is more important. If an area is so much important, then why not scan the whole area at a time. The thermal lamp will not start from one end of a layer and reach to other end in order to follow lineswise scanning; but the lamp will just appear facing the whole area then will scan and disappear; the lamp is over a powder bed, the lamp is switched on and then off, the whole area is scanned instantly without any physical movement of the lamp, there is no longer any movement of the lamp from one side to the other as it happened in linewise scanning. This scanning covering whole area instantly or this areawise scanning will be faster than linewise scanning. Thus, areawise scanning is convenient, but it has no power of judgement, it cannot distinguish which area needs to be consolidated; if all scanned area will be solidified then there will always be a rectangular block and there would not be any useful part. Areawise scanning needs help for making a useful part. Areawise scanning will be helped if thermal radiation coming from a lamp is prevented from reaching each and every part of an area. This can be done by using a physical mask between the lamp and the powder bed so that the physical mask will prevent radiation from the lamp to reach to the powder bed; therefore, though the whole area is scanned, the effect of thermal radiation will not be materialized below the masked area (Hermann and Larson 2008). Areawise scanning can also be helped if instead of creating a physical mask, a mask-type mechanism can be created within the lamp so that when the lamp is switched on, the radiation will not emanate from the whole area of the lamp but only from the unmasked area of the lamp (Farsari et al. 2000). If radiation will come from the selected area of the lamp, then it will affect only the selected area of the powder bed, and thus the selected area will be solidified, and thus useful part will be formed. Areawise scanning can also be helped if instead of using any type of masks, powders of selected area of a powder bed will be changed so that the selected area will act differently when all area will be radiated. If a selected area will be mixed with radiation absorbing material (same in HSS) then that area will be solidified and thus helping to make useful parts (Ellis et al. 2014). If the selected area will have higher melting point material then that will not be solidified but other non-selected area will be solidified and thus a useful part will be formed (Khoshnevis et al. 2014). An example of areawise scanning- microheater array was set above powder bed in microheater array powder sintering (MAPS) (Holt et al. 2018).

Areawise scanning has also been called layerwise scanning (Gibson et al. 2010; Holt et al. 2018). But layerwise scanning also implies layer by layer scanning and has been used as such for long (Deckard et al. 1992). It will be thus difficult to distinguish between two meanings of layerwise scanning which are as follows: (1) scanning a layer and completing it and going to the next layer and again scanning, and (2) scanning each and every point of the layer at a time. The name areawise scanning unlike the name layerwise scanning has no such precedence and is free from such ambiguity.

5.4.4 Basic Difference Between the Three Types of Scanning

If area of a powder bed is big and exposure area of a lamp is small, then whole area of the powder bed will not be scanned by just switching on and off the lamp. The lamp needs to move to cover the whole area. If area of a powder bed is very big and exposure area of a lamp is very small, then the lamp needs to move not only in one direction but also in other directions to cover the whole area. This brings several questions: if exposure area of lamp is very small then the area will be called an area or a point, how small is very small so that an area will become a point, which point is not having an area, what if the point is in the form of a line, which line has no area; is there anything more than a physical dimension (point, line, area) that sets these three types of scanning apart.

The purpose of these three types of scanning is to make an area. In pointwise scanning, an area is made if a point moves in a plane or if a point moves in both x- and y- directions in an xy plane; if a point moves only in a direction then it will end up making a line instead of an area; if the point does not move at all, then it will not even make a line, it will make just a point. In linewise scanning, the line needs to move in only one direction (normal to the line) (Zhu et al. 2000) to make an area; if a line does not move, it will not make an area but it will end up making just a line; if a line moves in a plane just like a point, i.e. moves in both x and y-directions, it will make bigger area, but for making just an area, movement in one direction is sufficient. In areawise scanning, without any movement of scanner or lamp, an area can be scanned or, scanning of an area can be happened.

In summary, in pointwise scanning, there is a minimum requirement of movement in two directions or two dimensions to scan an area; in linewise scanning, there is a minimum requirement of movement in one direction or one dimension to scan an area; in areawise scanning, there is a minimum requirement of movement in zero direction or zero dimension to scan an area. These are the basic differences among three types of scanning. Thus, if an area of size 500 micron × 500 micron is scanned by exposing this area once with a point source of area of 500 micron × 500 micron then it is an areawise scanning rather than a pointwise scanning. In another example, if an area of size 5 m × 5 m is scanned by moving five times in sequence a big lamp of area of 1 m × 5 m then this type of scanning is a linewise scanning rather than an areawise scanning. Figure 5.3 shows schematic diagram of three types of scanning: in Fig. 5.3a a laser beam is used to scan a powder bed, in Fig. 5.3b a number of laser beam emitters are fixed in a line, in Fig. 5.3c an array of heat sources covers whole powder bed.

Analogy with daily life: pointwise scanning – a bird picking grains in a field – hard work; linewise scanning – swiping a credit card in supermarket – fast work; areawise scanning – a person protecting oneself using umbrella in a rainy season – work without movement.

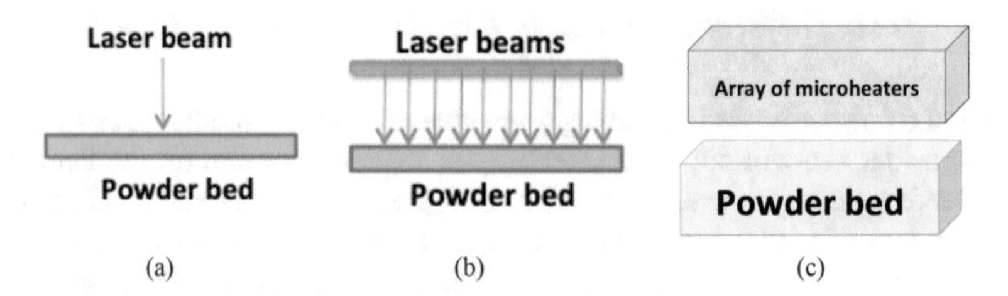

Fig. 5.3 Schematic diagram of pointwise, linewise and areawise scanning: (**a**) pointwise scanning, (**b**) Linewise scanning (**c**) Areawise scanning

5.5 Binder Jet Three-Dimensional Printing

Binder jet three-dimensional printing (BJ3DP) is a powder bed non-fusion process in which binder from an ink jet printhead is used to join powders (Fig. 5.4). The process is similar to beam based powder bed processes except that the job of beam is done by a binder jet. This process is also called 3DP but the name 3DP has potential to be confused with general 3DP, a synonym for AM, and therefore BJ3DP is used instead (Enneti et al. 2018; Kernan et al. 2007).

Since binder does not melt powders, or powders are joined without going through any type of melting; a number of powders irrespective of their melting points can be processed. Since binders are usually not the constituent of a final part, they need to be separated from the final part without disturbing the shape and dimension of the part. This requires post-processing with an aim to fulfil the following: (1) removal of binder so that a part will not continue to remain impure with binder, (2) solid state sintering so that the structure of the part will not change, (3) infiltration so that density and strength can be enhanced. Requirement of post-processing says that BJ3DP is not like other AM processes that give final products without such post-processing.

Role of binder is given below.

5.5.1 *Role of Binder*

Binder is jetted on powder bed, jetting implies that binder could be either liquid or air. There is no air based binder though there is an air deposition based technique named as aerosol jetting available, thus all jetted materials are liquid. Binder comes from a nozzle of an ink jet printhead – whatever jetted from the nozzle is supposed to act as a binder (to join) or to act as a partial binder (to cause to join). Binder means it is sufficient to bind powders without seeking an active contribution from the powder itself, e.g. joining of iron or copper powders by an organic binder. The binder joins iron powders as much efficiently as it joins copper powders; if there is a

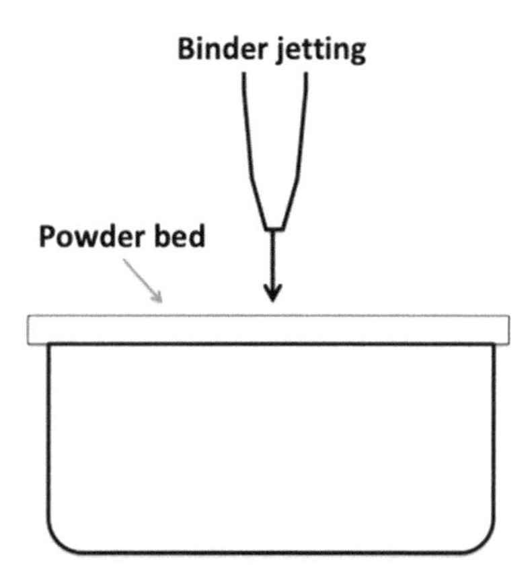

Fig. 5.4 Schematic diagram of binder jet three-dimensional printing (BJ3DP)

difference in joining strength, then it is due to difference in size or surface texture of the powder, but the difference is not because iron as a material acts differently than what copper does. Consequently, powder is neutral and does not contribute to increase the capability of binders, the situation will remain same if iron or copper will be replaced with aluminum or tungsten.

But what will happen if powder is not neutral – powder may be made of binders, powder is itself binder, powder is admixtured with few percentages of binder, or powder plays different roles and one of its role is to act as a binder in the presence of certain materials. In these cases, whatever is coming out from the nozzle may not completely qualify to be called as a binder, in utter disregard to the name of the process. But, this jetted material may also not completely disqualify as a binder because if this material does not trigger a reaction, does not cause a binding to happen on the powder bed then how the nozzle will make pattern on the bed, how the nozzle will continuously help create a 3D shape out of plane deposited layers. This jetted material may not be called binder but may not be deprived to be called as a partial binder. An example of this type of jetted materials is using water as a binder on cement powders. Water reacts with cement and binds cement powders; the same water may not be used as a binder for iron powders.

The use of partial binder is following: (1) binder solely coming from a nozzle may not be able to go through the thickness of the layer; if some binder is present in the layer, it will make up the binding process; (2) binder clogs the nozzle, a diluted binder may not, (3) a diluted binder may improve rheology of liquid jet, may help form better drop and improve printing process; (4) a better binder which may not leave residue after post-processing; (5) binder may not be required, such as water on cement or plaster; (6) jetted material is not intended to be removed during post-processing but is required as a constituent of a final part.

The ultimate aim of jetted material is to bind those zones and areas of the powder bed where it is jetted. But what if it does not bind only those areas where it is jetted – in this case, an object looking like negative 3D image of a 3D object will be formed. This process is called selective inhibition sintering (SIS) – as the name implies, jetted material inhibits binding to take place (Khoshnevis et al. 2014). This is possible if jetted area has lower ability to bind while other area is full of binder so that during post-processing heat treatment, other area will bind while jetted area will not bind – jetted area is immune to high temperature.

The main difference between BJ3DP and SIS:

1. If a big part is to be made then, in BJ3DP, it will require a lot of binders to be jetted while in SIS, it will require material to be jetted only on boundaries. Thus using SIS than BJ3DP, more binder will be saved.
2. If a complex part is to be made then in BJ3DP, the complex part bound with binder (known as green part) will be taken to furnace for post-processing. In case of SIS, as usual whole powder container containing jetted area and surrounding powder will be taken to furnace for post-processing. In most geometries, surrounding powder only on one side of the jetted line or jetted area is intended to become part while that of other side will become waste. Thus, in SIS, the powder which will not contribute to the final part will also be subjected to furnace treatment and will become unusable. Thus, using BJ3DP than SIS, more powder will be saved.

Figure 5.5 shows the difference between BJ3DP and SIS. Figure 5.5a shows binder deposition in BJ3DP, binder is deposited in yellow area, the part forms (Fig. 5.5b) as per deposited area. Figure 5.5c inhibiter deposition in yellow area, the part forms (Fig. 5.5d) from other area which is surrounding powder (red in color). Outside the yellow area, the part may not be used and will become waste. Using a material

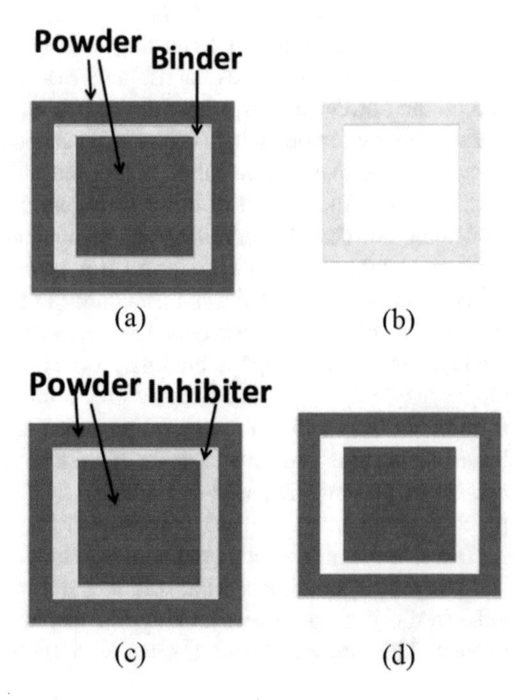

Fig. 5.5 Role of binder and inhibiter in BJ3DP and SIS: (**a**) binder jetting, (**b**) final part in BJ3DP, (**c**) inhibiter jetting, (**d**) final part in SIS

(inhibiter) to cause non-adding in additive manufacturing is not unique to SIS, anti-glue (Gibson et al. 2010) or carbon powder (Liao et al. 2003) is used in sheet based process to cause selective non-joining between two sheets.

References

Arredondo MZ, Boone N, Willmott J et al (2017) Laser diode area melting for high speed additive manufacturing of metallic components. Mater Des 117:305–315

Athreya SR, Kalaitzidou K, Das S (2010) Processing and characterization of a carbon black-filled electrically conductive Nylon-12 nanocomposite produced by selective laser sintering. Mater Sci Eng A 527(10–11):2637–2642

Baumers M, Tuck C, Hague R (2015) Selective heat sintering versus laser sintering: comparison of deposition rate, process energy consumption and cost performance. In: SFF Proceedings, pp 109–121

Brown R, Morgan C T, Majweski C E (2018) Not just nylon – improving the range of materials for high speed sintering. In: SFF Proceedings, 1487–1498

Buls S, Vleugels J, Hooreweder B V (2018) Microwave assisted selective laser melting of technical ceramics. In: SFF Proceedings, pp 2349–2357

Chatham CA, Long TE, Williams CB (2019) A review of the process physics and material screening methods for polymer powder bed fusion additive manufacturing. Progr Polym Sci 93:68–95

Dallarosa J, O'neill W, Sparkes M, Payne A (2016) Multiple beam additive manufacturing. Patent WO2016201309A1

Deckard C R, Beaman J J, Darrah J F (1992) Method for selective laser sintering with layerwise cross-scanning. US5155324A

Ellis A, Noble CJ, Hopkinson N (2014) High Speed Sintering: assessing the influence of print density on microstructure and mechanical properties of nylon parts. Addit Manuf 1–4:48–51

Enneti RK, Prough KC, Wolfe TA et al (2018) Sintering of WC-12%Co processed by binder jet 3D printing (BJ3DP) technology. Int J Refract Met Hard Mater 71:28–35

Farsari M, Claret-Tournier F, Huang S et al (2000) A novel high-accuracy microstereolithography method employing an adaptive electro-optic mask. J Mater Process Technol 107(1–3):167–172

Fredriksson C (2019) Sustainability of metal powder additive manufacturing. Proc Manuf 33:139–144

Gibson I, Rosen DW, Stucker B (2010) Additive manufacturing technologies: rapid prototyping to direct digital manufacturing. Springer, New York

Goodridge RD, Shofner ML, Hague RJM et al (2011) Processing of a Polyamide-12/carbon nanofibre composite by laser sintering. Polym Test 30(1):94–100

Hagen D, Kovar D, Beaman JJ (2018) Effects of electric field on selective laser sintering of yttria-stabilized zirconia ceramic powder. In: SFF Symposium Proceedings, pp 909–913

Hecht J (2018) Understanding lasers: an entry-level guide. Wiley, Hoboken

Hermann D S, Larson R (2008) Selective mask sintering for rapid production of parts, implemented by digital printing of optical toner masks. In: NIP & digital fabrication conference

Ho HCH, Cheung WL, Gibson I (2002) Effect of graphite powder on the laser sintering behaviour of polycarbonate. Rapid Prototyp J 8(4):233–242

Holt N, Horn AV, Montazeri M, Zhou W (2018) Microheater array powder sintering: a novel additive manufacturing process. J Manuf Process 31:536–551

Hong R, Zhao Z, Leng J et al (2019) Two-step approach based on selective laser sintering for high performance carbon black/polyamide 12 composite with 3D segregated conductive network. Compos Part B: Eng 176:107214

Jerby E, Meir Y, Salzberg A et al (2015) Incremental metal-powder solidification by localized microwave-heating and its potential for additive manufacturing. Additive Manufac 6:53–66.

Kernan BD, Sachs EM, Oliveira MA, Cima MJ (2007) Three dimensional printing of tungsten carbide-10 wt % cobalt using a cobalt oxide precursor. Int J Refract Met Hard Mater 25:82–94
Khoshnevis B, Zhang J, Fateri M, Xiao Z (2014) Ceramics 3D printing by selective inhibition sintering. In: SFF proceedings, pp 163–169
Kumar S, Czekanski A (2018) Roadmap to sustainable plastic additive manufacturing. Mater Today Commun 15:109–113
Leonelli C, Veronesi P, Denti L et al (2008) Microwave assisted sintering of green metal parts. J Mater Process Technol 205(1–3):489–496
Li L (2000) The advances and characteristics of high-power diode laser materials processing. Opt Lasers Eng 34(4–6):231–253
Liao YS, Chiu LC, Chiu YY (2003) A new approach of online waste removal process for laminated object manufacturing (LOM). J Mater Process Technol 140(1–3):136–140
Pinkerton AJ (2016) Lasers in additive manufacturing. Opt Laser Technol 78A:25–32
Salehi M, Maleksaeedi S, Nai MLS, Gupta M (2019) Towards additive manufacturing of magnesium alloys through integration of binderless 3D printing and rapid microwave sintering. Addit Manuf 29:100790
Santos ES, Shiomi M, Osakada K, Laoui T (2006) Rapid manufacturing of metal components by laser forming. Int J Mach Tools Manuf 46(12–13):1459–1468
Sillani F, Kleijnen RG, Vetterli M et al (2019) Selective laser sintering and multi jet fusion: process-induced modification of the raw materials and analyses of parts performance. Addit Manuf 27:32–41
Thomas HR, Hopkinson N, Erasenthiran P (2006) High speed sintering – continuing research into a new rapid manufacturing process. In: SFF proceedings, pp 682–691
Tian Y, Tomus D, Rometsch P, Wu X (2017) Influences of processing parameters on surface roughness of Hastelloy X produced by selective laser melting. Addit Manuf 13:103–112
Wagner T, Hofer T, Knies S et al (2005) Laser sintering of high temperature resistant polymers with carbon black additives. Int Polym Process 19(4):395–401
Zhu L, Cheng J, Zhou H (2000) Research of rapid prototyping process using linear array of high power laser diodes. In high power lasers in manufacturing, In: Proceeding of SPIE 3888

Chapter 6
Beam Based Solid Deposition Process

Abstract Beam based solid deposition process is used to fabricate large parts having medium complexity. This chapter describes laser beam based and electron beam based solid deposition process. Different types of powder depositions such as coaxial continuous, coaxial discrete and off-axial are given while laser-powder interactions are briefly explained. How repair occurs is explained with an example while difference between two types of feedstocks such as wire and powder in terms of their efficiencies in processing is elaborated.

Keywords Wire · Laser-powder interaction · Electron beam · Powder · Repair

6.1 Introduction

Solid deposition process (SDP) is a family of additive manufacturing processes in which solid materials are used as feedstocks and brought as a solid to a point on a platform (or on a substrate) where they are converted in a desired shape. Feedstocks in the form of powder, wire and rod are used; and in order to transform the feedstock in a 3D part, energy is required. Energy comes either from beams such as laser, electron or from arc, or from heater, or from mechanical sources (friction or motion of the feedstock). Beam based processes are laser engineered net shaping (LENS) (Pinkerton 2016), electron beam additive manufacturing (EBAM) (Tarasov et al. 2019) etc.; arc based process is wire arc additive manufacturing (WAAM) (Tabernero et al. 2018; Cunningham et al. 2018) and plasma welding based additive manufacturing (Feng et al. 2018); friction based process is additive friction stir deposition (AFSD) (Yu et al. 2018); heater based process is fused deposition modeling (FDM) (Masood 2014); the process which uses its own motion as energy source is cold spray additive manufacturing (CSAM) (Yin et al. 2018).

On the basis of energy sources used, solid deposition process can be divided into two broad categories: beam based solid deposition process and non-beam based solid deposition process. Beam based solid deposition process can be further divided into three categories: laser beam based, electron beam based and plasma-beam based. Non-beam based solid deposition process can be divided into four

S. Kumar, *Additive Manufacturing Processes*,
https://doi.org/10.1007/978-3-030-45089-2_6

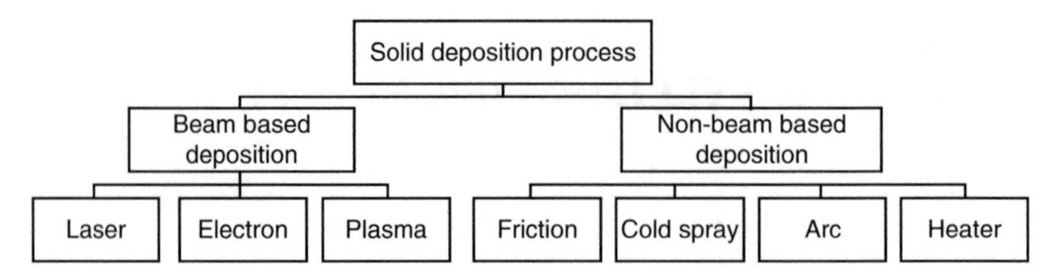

Fig. 6.1 Classification of solid deposition process

categories: friction based, cold spray based, arc welding based and heater based, as shown in Fig. 6.1.

This chapter deals with beam based solid deposition process while non-beam based solid deposition process is dealt in Chap. 7. Since plasma beam used in plasma beam based solid deposition process is solely generated from arc used in arc welding; plasma beam based SDP is described along with arc welding based SDP in Chap. 7. Thus, this chapter deals with laser beam based SDP or laser SDP (LSDP) and electron beam based. LSDP is used both for powder based LSDP and wire based LSDP. Electron beam based SDP is named as electron beam additive manufacturing (EBAM) that uses only wire as a feedstock.

6.2 Laser Solid Deposition Process

Laser solid deposition process (LSDP) has many names such as directed energy deposition (DED), laser engineered net shaping (LENS), laser powder deposition (Vilar 2014), laser based additive manufacturing (Yan et al. 2018) etc. The process akin to other AM processes starts with a CAD file. A laser beam coming from a nozzle acts as a heat source; feedstock (blown powder or fed wire) supplied by either same nozzle or different nozzle is melted by the beam at a point of interaction on the substrate. By moving the nozzle parallel to a fixed substrate or by moving the substrate parallel to a fixed nozzle, the point becomes a line; by creating such lines as per the design of the first layer of a CAD file, the first physical layer is fabricated. By moving the nozzle up or the substrate down, space is created to accommodate second to-be-fabricated layer; the step of nozzle movement and melting is repeated to fabricate the second layer. Successive layers of the CAD file are made in the same way (Schmidt et al. 2017). The process has a number of variations, which is due to variations in a nozzle, feedstocks, position of the nozzle, laser-material interaction etc. This process is an extension of laser cladding. There is a subtle difference between laser cladding and this process. Laser cladding is meant to apply coating on a substrate in order to protect it or to improve its properties, the coating usually conforms to the contour of the substrate. While LSDP makes a three dimensional structure on a substrate which may be cut off from the substrate to be further used, in laser cladding the coating is an integral part of the end part and the coating is not cut off from the substrate to be used separately. During repair and refurbishment, LSDP makes additional features on a substrate. If LSDP akin to laser cladding makes coating on a substrate conforming to its geometry, then it is a case of laser cladding process executed by an LSDP system.

6.2.1 Types of Powder Deposition

Three types of powder deposition are generally practised in LSDP. These are given below.

6.2.1.1 Coaxial Continuous Powder Deposition

In this type, powder is deposited around the axis of a laser beam (Fig. 6.2a), the beam comes from inside a nozzle while powders come from a continuous annular orifice (Fig. 6.2d) at the periphery of the nozzle. The powder stream takes conical shape while enclosing the beam; the axis of the cone is same as that of the laser beam, giving the name coaxial to this type of deposition. Since the powder is injected from all points at the annular space, this coaxial powder deposition is called continuous. This type of deposition is used for fabricating structures with high resolution (Oliveira et al. 2005). Since the deposition has no directional dependence, it is suitable to make multi-layer deposition (Eisenbarth et al. 2019).

6.2.1.2 Coaxial Discrete Powder Deposition

In this type, powder is deposited not from all points at the annular space as happens in the former type but from certain selected points (Fig. 6.2b). Figure 6.2e shows three discrete points from where powder is injected forming a conical shape (having same axis as that of the beam) around the beam. For those cases having less than three discrete points, a conical shape will not form excluding those cases from this type of deposition. Discretization helps achieve higher deposition rate but it has lower symmetricity of powders' position around the beam in comparison to the former type; increasing the number of discrete points will increase the number of powder jets and thus the symmetricity. Using separate powder nozzles for every powder jet will also provide the same effect.

6.2.1.3 Off-Axial Powder Deposition

In this type, the axis of a powder stream makes an angle more than 0° and less than 90° with a laser beam axis (Fig. 6.2c). At 90°, a powder jet will not impinge upon a substrate unless either the substrate is tilted or a powder jet loses its speed and falls

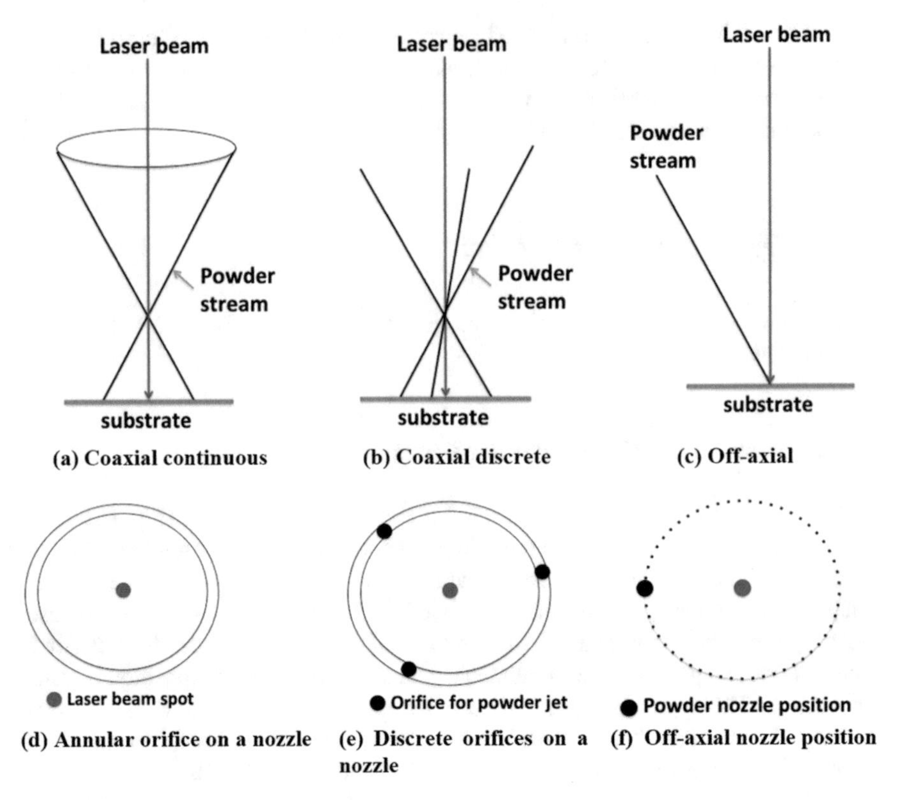

Fig. 6.2 Schematic diagrams for powder deposition: (**a**) coaxial continuous powder deposition, (**b**) coaxial discrete powder deposition, (**c**) off-axial powder deposition, (**d**) top view of annular orifice on a nozzle, (**e**) top view of discrete orifices on a nozzle, (**f**) top view of off-axial powder nozzle position

on a substrate due to gravity. At 0°, the powder jet will no longer be off-axial but will become coaxial. The nozzle for powder jet is generally not associated with the nozzle for the laser beam which gives flexibility in setting up an off-axis angle and a stand-off distance of the powder nozzle from the substrate. This type of deposition is used for its simplicity and flexibility (Vilar 1999). However, this has directional dependence, which implies that due to the motion of the nozzle (or the substrate), the amount of the powder to be deposited and the intended place for the deposition on a substrate will change (Fig. 6.3). The powder stream does not actually follow the straight line as shown in a schematic diagram (Fig. 6.2c) but is bent due to gravity as shown in Fig. 6.3a, b. Powders after leaving the nozzle are no longer tied to the nozzle and are flowing to reach to the substrate; when the substrate is moving towards them, powders strike the substrate at a nearer location (Fig. 6.3a); when the substrate is moving away from them, powders will strike at a farther location (Fig. 6.3b); this causes directional dependence as due to the change in direction of motion, intended strike position will be negatively or positively missed. This situation can be avoided in following two cases: (1) powder jet is injected with a higher

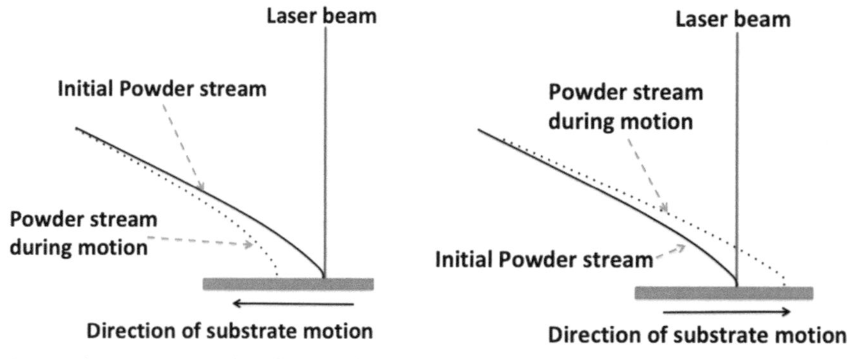

(a) Powder stream and substrate have opposite direction

(b) Powder stream and substrate have same direction

Fig. 6.3 Directional dependence in off-axial powder deposition: (**a**) powder stream does not reach to laser-substrate interaction point, (**b**) powder stream goes beyond laser-substrate interaction point

speed so that it will not be affected by gravity but it will cause stronger ricochet from a solid substrate, and cause ripples in the melt pool, (2) the substrate moves with a slower speed (of the order of mm/s, slower than the powder feed rate), it will decrease the production rate, (3) powder nozzle should be kept closer to the substrate, though it may damage the nozzle. If the time for powder to reach the substrate is small or of the order of millisecond, it will not be affected by gravity or drag force (Wu et al. 2018).

Implications of directional dependence depends on other process parameters; such as, if a laser beam spot size is far smaller than a powder stream spot size, and with displacement of the powder stream (due to any motion), the laser beam spot size is always covered by the powder stream spot size then there will be no implication. However, if due to motion, the laser beam spot is either deprived of or overflowed with powders, it will cost uniformity of the layer. Employing two powder jets in opposite directions will mitigate the effect of direction; in one direction one jet will cause deprivation while other will cause overflowing while in other direction reverse will happen; consequently, the net effect will be zero. Figure 6.4b shows the covering of laser beam spot due to contribution from both powder streams during no motion of the substrate. During motion of the substrate in the left, right powder stream covers the laser beam spot, while during motion in the right (Fig. 6.4c), left powder stream covers the spot (Fig. 6.4d).

6.2.2 *Laser-Powder Interaction*

In this process, the timing of interaction between a laser beam and a powder stream influences the process, there are three possibilities; the process can be any type of combination of these possibilities. These three are as follows:

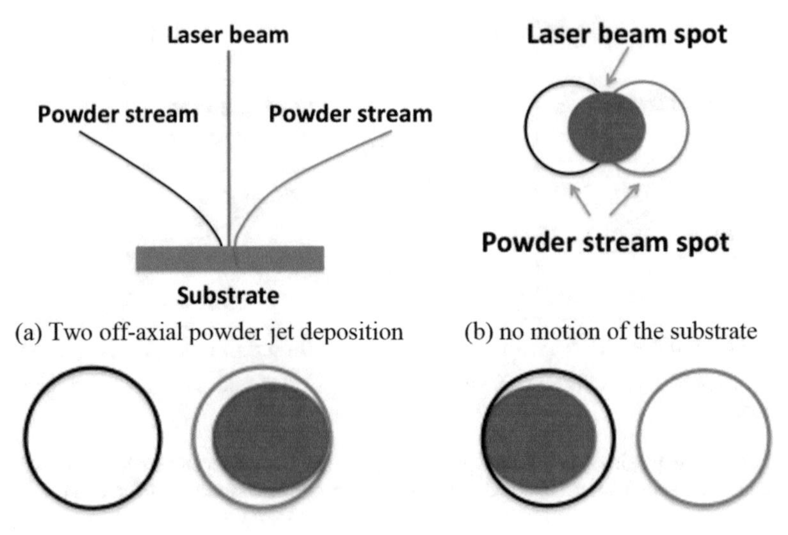

(a) Two off-axial powder jet deposition (b) no motion of the substrate

(c) substrate motion in the left direction (d) substrate motion in the right direction

Fig. 6.4 Directional dependence in two off-axial powder jet deposition: (**a**) schematic diagram of two powder jets deposition, (**b**) interaction between laser beam spot and two powder stream spots, (**c**) displacement of powder stream spots in the left direction due to the substrate motion in the left direction, (**d**) displacement due to the substrate motion in the right direction

6.2.2.1 Laser Beam and Powder Stream Interacts Midway

If a laser beam and a powder stream meet on the way before they reach a substrate, it can affect the process. If powder comes on the way, it will obstruct the laser beam from reaching to the substrate, powder will be heated and the heated powder will reach to the substrate, this shadowing effect will bring variation in melt pool. In case of coaxial continuous powder deposition, the degree of interaction depends upon the geometry of laser beam cone and the cone made by powder stream (Li and Huang 2018) as shown in Fig. 6.5. In case the laser beam cone is thinner than the surrounded powder stream envelope, then the beam will intersect the envelope in smaller zone (Fig. 6.5a), otherwise the intersection zone will be bigger (Fig. 6.5b). Bigger intersection zone will cause bigger obstruction of the beam which will cause heating of bigger amounts of powders. The heating of powders can be minimized or avoided by using an off-axial powder deposition for depositing powder either at an angle or perpendicular to the substrate in combination with an inclined laser beam as shown in Fig. 6.6a, b, respectively. It can also be minimized in a usual setup as given in Fig. 6.2c or Fig. 6.6c. Inclined laser beam (Meacock and Vilar 2008) though will increase the dissipation of laser energy by increased reflection of the beam. Increasing the angle between laser beam path and powder stream path as shown in Fig. 6.6a will have minimum shadowing effect but this will decrease both energy efficiency and powder catchment efficiency (this is a percentage of blown powders ended up in a clad or part).

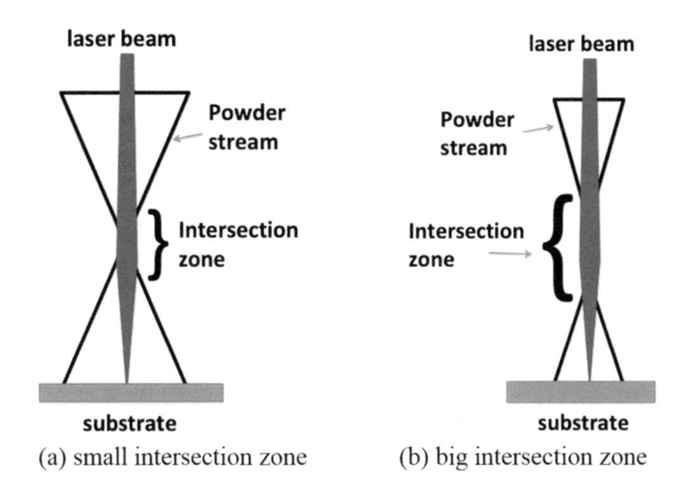

Fig. 6.5 Intersection between laser beam and powder stream in coaxial continuous powder deposition: (**a**) small laser-powder intersection zone, (**b**) big laser-powder intersection zone

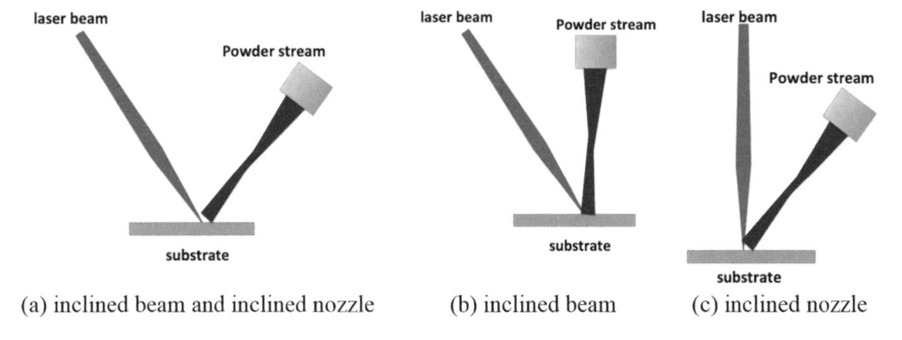

Fig. 6.6 Minimizing shadowing effect in off-axial powder deposition: (**a**) inclined laser beam and inclined off-axial nozzle, (**b**) inclined beam and perpendicular nozzle, (**c**) perpendicular beam and inclined nozzle

If the powders, instead of getting just heated, get melted then the creation of melt pool on the substrate will no longer be required as a means to melt powders and the problem associated with creating and controlling the melt pool will be over. There are the following reasons why powders are not melted: (a) powders are not small enough; powders used in this process are from the size of about 20 micron and above, these powders are not small enough to be melted in such small laser-powder interaction time. If these powders could be of sub-micron size, they can melt but using sub-micron or nano-sized powder is not possible as these cannot be fluidized (Geldart 1973) and accelerated using carrier gas. During acceleration, they agglomerate and no longer remain an isolated small powder small enough to be melted. If powders are very small, they will vaporise and can form plasma which will attenuate laser beam. If powders are not very small, then they will partially vaporise which will give rise to recoil pressure on the powder and which will change the direction of the flow of the powder (Sergachev et al. 2020). Which powder is small and which one is very small depends upon experimental conditions. (b) Laser energy is not sufficient – focal length of laser beam is small which does not let the powder to have sufficient interaction with the beam to gain sufficient laser energy for melting; this can be overcome by increasing the focal length as well as increasing the laser power. Increasing the focal length to a big length, though possible by optics, will bring an additional problem to have a big focussed powder stream matching that big length. Increasing the laser power may vaporise the substrate in the course of melting the powder unless the laser is diverted by additional optics. Thus, the problem can be overcome and liquid drops (by melting powders) will fall on the substrate, which on being continued will give rise to a 3D structure. But, overcoming this problem will bring a question mark on an effort for overcoming this problem; the aim of the process is not to facilitate complete melting on the way but the opposite – to prevent even an isolated case of powder melting on the way; besides there are more economic and energy-efficient ways to create liquid drops than using expensive powders and laser.

6.2.2.2 Laser Beam Follows Powder Stream

Powders are blown on a solid substrate in order to deposit them; powders will be rebounded from the substrate and the substrate will not be having the same amounts of powders as intended. With an increase in feed rate and size of the powder, rebound force (or repelling force) (McLaskey and Glaser 2010) will increase, it will make increasingly difficult to place a layer of powder on the substrate and consequently to replicate a powder deposition type familiar either in powder bed fusion or in preplaced version of laser cladding. This difficulty is also a cause or source of difference between two processes – powder bed type and powder blown type. Imagining a state where this difficulty disappears – the nozzle will be used to additionally create a powder layer which will act as a support to make overhangs on it and there will not be any need to manoeuvre the axes of a 5-axis CNC machine to circumvent the need of support structure – this will, if not make powder bed type process extinct, certainly take away the edge powder bed type has over powder blown type. Though, the difficulty will not disappear but can be surmounted with the help of an additional nozzle fitted with the sole purpose of placing powders (details given in Chap. 12).

In this process, though it is impossible to make a powder layer, it is also impossible at the same time not to have a single powder on the substrate when powders are blown on the substrate. These few powders, somehow survived on the substrate, count when the aim is to create a thin melt pool (of μm thickness), these few powders will take the heat away from the substrate, and prevent the substrate from getting overmelted. A thin melt pool is desired when minimum dilution (it is a measure of how much substrate material is becoming part of the build material) is required. Besides, creating a thin melt pool rather than a thick melt pool will also need lower energy resulting in lower heat accumulation causing lower raise in temperature of the substrate. During processing of some materials, when temperature is rising and reaching to a critical point, starting with a substrate having low temperature will allow to work for longer duration before a critical point is reached. Working for longer duration means attaining a higher height of the built; in case of epitaxial material, it means gaining a higher height of desired columnar grain before it becomes equiaxed (Kirka et al. 2009). A thin melt pool is always desired when a thin melt pool can serve the purpose even if the purpose is not a special purpose, because it always requires lower energy to create a thinner pool. There can be a number of strategies (such as decreasing laser power, increasing scan speed) to create a thin pool, and these strategies can be better than how a thin melt pool is just resulted due to the presence of stray powders. The point is not which strategy is better but the point is what will be the consequence if laser beam happens to reach the substrate after powder has already reached.

6.2.2.3 Powder Stream Follows Laser Beam

When a laser beam strikes a substrate, it is supposed to create a melt pool so that when powder stream reaches to the substrate, powders will be melted in the melt pool and the deposition of material will start. How long these powders remain unmelted in the melt pool determines the surface roughness and build quality; if they remain unmelted for long, then there will be no more space for other powders to be accommodated in the melt pool (Haley et al. 2019); if incoming powders cannot be accommodated, then the build rate will decrease. Though, powders attenuate the coming beam depending on the configuration of laser beam and powder stream, this type of creation of melt pool and subsequent deposition is practised in multilayer deposition. From a powder stream how many powders melt pool is going to catch determines the efficiency of powder deposition – this is called catchment efficiency. Size of the melt pool will increase with an increase in injection of powders and its temperature will increase with increase in injection of hot powders (Pirch et al. 2019).

6.3 Repair in Laser Solid Deposition Process

A part having damage on its surface such as wear, scratch, corrosion, dent or crack can be repaired by laser beam deposition. In case of cracks on a surface, coating or filling up cracks will not work as cracks can still propagate from the crack tip, besides the bonding between filled-up material and cracked surface will not be strong enough to last. All cracks need to be removed from the part for effective repair. Cracks, which are bigger and cannot be removed without removing excess material, cannot be repaired because it will not be cost effective. Besides, the laser beam cannot access the repair site or removing the material around the repair site

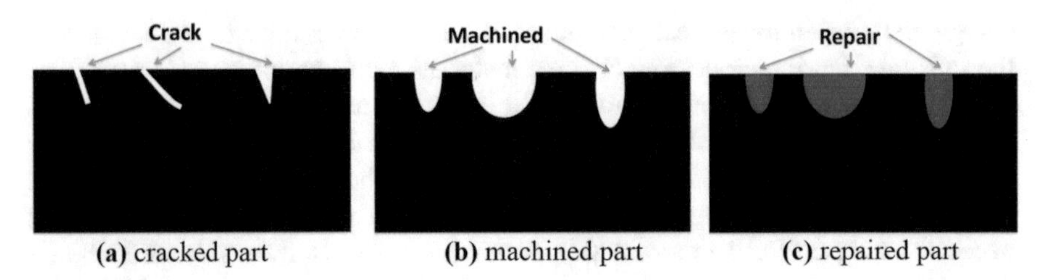

Fig. 6.7 Repair of cracks in LSDP: (**a**) parts having cracks, (**b**) cracks are machined, (**c**) cracks are filled up

may not be possible. Thin hair cracks reaching to the central area or cracks located at the central area are not suitable for repair. Machining may require evaluation of the damaged surface by modelling (using reverse engineering), evaluating the optimum amount of materials to be machined and generation of toolpath (Zhang et al. 2018). Machining more than optimum materials may neither cause technical problems in repair nor will result in adverse material properties, but is against the basic principle of AM that not to use machining (and generate chips) when machining is not required. The lofty goal of AM that not to use machining (and generate chips) even when machining is required will not be achieved in this case.

Figure 6.7 shows the stage for repair; Fig. 6.7a shows a part having three cracks which are subsequently machined (Fig. 6.7b) and repaired (Fig. 6.7c) by LSDP. Repair gives an opportunity to use better material than the original material of the parts and improve the properties not achieved by original undamaged parts. With an advent of LSDP, crack is no longer a problem but an opportunity for betterment.

6.4 Electron Beam Additive Manufacturing

Electron beam additive manufacturing (EBAM) is a process in which a 3D part is made by feeding wire in a melt pool created by an electron beam. The movement of substrate in x and y direction with respect to an electron beam makes a layer while layer addition takes place in z direction as shown in Fig. 6.8. Electron beam loses energy when it ionizes gas molecules present in the environment. In order to prevent ionization, electron beam needs to travel in vacuum. Therefore, interaction between wire, substrate and electron beam takes place in a vacuum chamber (Fig. 6.8). The need of a vacuum chamber limits the size of the part that could be made and thus this process is not having as much freedom as laser solid deposition process (LSDP) does have; since, LSDP does not need a vacuum chamber. Presence of vacuum chamber not only limits the size of the part (that could be fabricated and repaired) but also limits the type of feedstock that could be used. In a vacuum chamber, deposition of powder will not be convenient – powder may ionize on the way, small powders may fly because of charging, carrier gas which is bringing powder to the substrate will ionize as well. Therefore, EBAM is associated with the wire as feedstock.

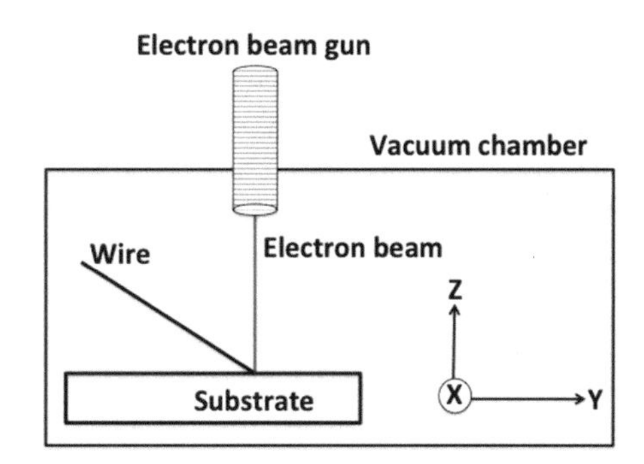

Fig. 6.8 Schematic diagram of electron beam additive manufacturing

EBAM is an energy-efficient process because feedstock is efficient (maximum portion of the wire fed ends up in deposition) and the energy source is efficient (95% electrical energy is converted into e-beam). There is a limit to the maximum power of e-beam that can be used because at high power, the x-ray generated due to e-beam needs to be shielded from coming out of the chamber.

6.5 Difference Between Powder and Wire as Feedstock

Both powder and wire are used as feedstock in almost all solid material deposition processes. The difference between both is as follows.

6.5.1 Cost

Powder is more expensive than wire. Many times, powder is produced from wire and therefore the cost of powder is cost of wire plus cost of conversion of wire into powder. Higher cost can also be perceived from the fact that for the same material getting narrow distribution of powder is more difficult than getting uniform diameter of wire. Powder requires carrier gas to be delivered, which increases the total cost. Low cost of wire gives advantage when a big part is fabricated (Hassen et al. 2020).

6.5.2 Availability

Wire of limited materials are available, the available wire is mostly made from metals – the name wire is traditionally used for metallic wire. If the metal or alloy is malleable, it is easy to draw wire from such materials and it increases the availability. Wires from plastic and ceramics are available but these are not used in

SDP. Powder has higher availability than wire because powder can be made from a higher number of techniques – from solution (joining atom by atom) as well as from big blocks (melting and spraying). Besides processing two or three types of powders (metal and ceramics), a wide variety of composite powders (usable in SDP) can be formed, while for processing two or more materials (metal and ceramic), a limited variety of composite wire (usable in SDP) can be formed. This is because wire needs to be drawn out while powder has no such compulsion, though, majority of the powders should have regular size. If there is irregularity in the diameter of a wire, whole wire can be discarded while if some powders are irregular, these irregular powders could be discarded and the remaining powders can still be used.

6.5.3 Material Efficiency

During deposition, there are many powders to control while there is just a single wire; consequently the chance of losing powders is higher while there is no such possibility with wire. Thus, powder gives lower deposition efficiency than wire (Schmidt et al. 2017). Powders left out during deposition needs to be collected, sorted and recycled while there is no such problem with wire, it increases the processing cost with powders. Besides, for creating the same size of part, higher amounts of powder are required. Nevertheless, powders furnish higher precision and accuracy than wire because powders can work with smaller melt pools while wire cannot work with small melt pool (comparable to its diameter) as it can partially melt and get entangled with the build causing termination of the process.

Wire is preferred in microgravity because whole wire can be traced, while some powder can spill and pose safety problems (Watson et al. 2002).

6.5.4 Processing in Vacuum

In case of EBAM, where processing is performed in vacuum, wire can be fed with ease without disturbing the vacuum. Feeding the powder using carrier gas changes the vacuum pressure due to the presence of gas, besides the gas gets ionized and obscures the electron beam which decreases the efficiency (Taminger and Hafley 2013).

6.5.5 Oxidation

For converting a material (powder or wire) into a bead, the material needs to be melted. For the same weight of material, material in the form of powder has higher surface area than the material in the form of wire. This implies during deposition or conversion or melting, if powder instead of wire is used, higher surface area of material is involved. Thus, in case of powder, higher surface area is exposed to oxidation in an oxidative environment – making powder based process more vulnerable to oxidation than wire based process.

6.5.6 *Effect on Process*

Powder gives flexibility to the process – a number of powders can be mixed and fed through a single nozzle to make a composite part while a number of wires cannot be fed from a single wire-feeder.

In wire based process, wire needs to be completely melted, but in case of variation of parameters, if the wire is partially melted the wire will be attached to the bead; it will lead to termination of the process. In powder based process, powder is expected to be completely melted, but in case of variation of parameters, if some powders are not melted, it can give rise to porosity but the process will still continue and will not be terminated. Feeding excess powder or insufficient powder can be a strategy for varying properties and making porous structures; wire feeding does not give such flexibility and variation in properties because wire feed rate needs to correspond with the heat input in order to prevent non-melting of the wire.

Wire is generally fed from sideways or an off-axial position, while powder can be fed both from sideways and coaxial position. Wire is localized at a certain point or a limited area in a melt pool, as shown in Fig. 6.9, while powders can be defocussed or focussed to cover all area of the melt pool. This implies that wire melts at a certain zone of the melt pool and spreads all over; how far the melted wire will spread depends upon how slow the melt solidifies. If it solidifies earlier, the surface of the melt pool would not be planar – it will decrease surface smoothness. While powder is injected at many places of the melt pool, though in smaller quantities in comparison to the bigger volume of the wire tip, melted powder will spread smaller distance or will not spread in comparison to the movement of melted wire. Consequently, surface smoothness will not be affected by melted powder as much as by melted wire – thus powder feeding gives better surface finish. However, all powders need to be melted, there are many powders in comparison to a single wire;

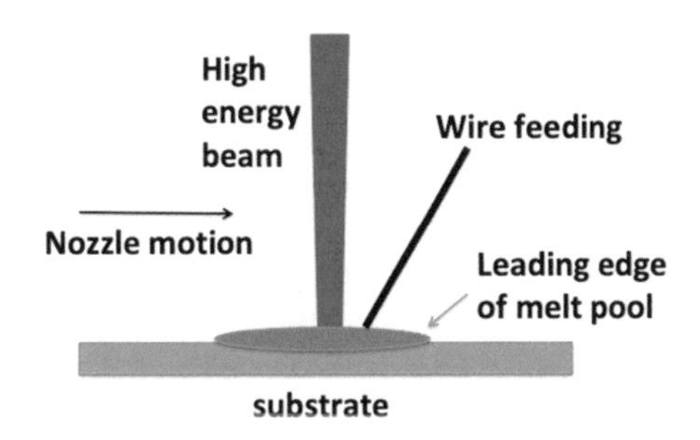

Fig. 6.9 Wire feeding at the leading edge of moving melt pool

with an increase in feed rate, it is more difficult to maintain confinement of all powders in melt pool than to maintain confinement of a single wire tip in the melt pool. Consequently, from the perspective of melting of feedstock, wire feeding rather than powder feeding allows to attain higher feed rate – thus higher deposition rate is achieved in case of wire.

Since wire feeding necessarily entails localization of wire tip at a particular place in the melt pool, it gives rise to a question – which place of the melt pool is a particular place of the melt pool. A melt pool is created by melting the substrate (or previously melted layer) using high energy beam, wire is positioned in such a way that it is going to be dipped in the coming melt pool. If the wire is positioned under the beam then it will attenuate the beam, therefore any position that is not under the beam can be the right place. For realization of the process, melt pool is created continuously by continuous movement of the high energy beam nozzle relative to the substrate (or movement of the substrate relative to the nozzle). A moving pool thus created has two edges, leading and trailing, since leading edge is in continuous contact with the beam therefore it is a hot edge while trailing edge is relatively cold and solidifies faster than the leading edge. Wire fed at the trailing edge has higher chance to solidify (and become part of the solidified melt pool) before it (the wire) acquires enough energy from the melt pool to melt itself (and let unmelted wire to march forward parallel to the substrate). This is the reason why the leading edge rather than the trailing edge is the preferred place of the melt pool where the wire is aimed to be positioned at (Fig. 6.9).

Direction of motion determines which edge of the melt pool is leading or trailing, during scanning if direction reverses the leading edge will become the trailing edge and vice versa. Therefore, if the position of the wire is not reversed with a reverse in direction, the performance of the process in both directions will not be same. Changing the position of the wire can be avoided if there are two wire feeders at either side of the beam, and with a change in direction alternate wire feeder will be used.

6.5.7 Processing an Inaccessible Area

Wire is a large integrated unit in comparison to disunited tiny powders – wire requires certain minimum stiffness to be pointed outside the wire feeder and to reach the melt pool, in absence of stiffness it will not move towards target but will fall due to gravity; this stiffness, which is an essential property of the wire, is also a problem when the aim is to reach inaccessible area to refurbish or repair. Powder unlike wire is not inseparable part of an integrated unit and the moment it leaves nozzle, it no longer remains attached to the nozzle, it has more possibility to reach an inaccessible point by other means such as rebounding and falling. Figure 6.10 shows an inaccessible area where at point A, material needs to be deposited. Vertical feeding is not possible due to the geometry of the structure, in case of vertical feeding if tip of the wire reaches point A then laser beam, instead of being able to enter the cavity, will be reflected away. By feeding wire from an off-axial position as shown in Fig. 6.11a, the point of repair, that is A, is not accessible. While in case of powder feeding at the same off-axial position, powder can still reach to point A by rebounding from the side surface and falling at A; this allows feeding to happen at the melt pool created by high energy beam (electron, plasma, laser) and build to be completed. In case of wire feeding, melt pool thus created by beam will not be fed by wire and the build will not be completed. This shows the powder as a feedstock has an advantage over wire as a feedstock. Though, in case of wire feeding deposition can still be accomplished by optimizing the parameters and melting the wire in the midway but the completion of deposition in this way again asserts the flexibility of powder feeding.

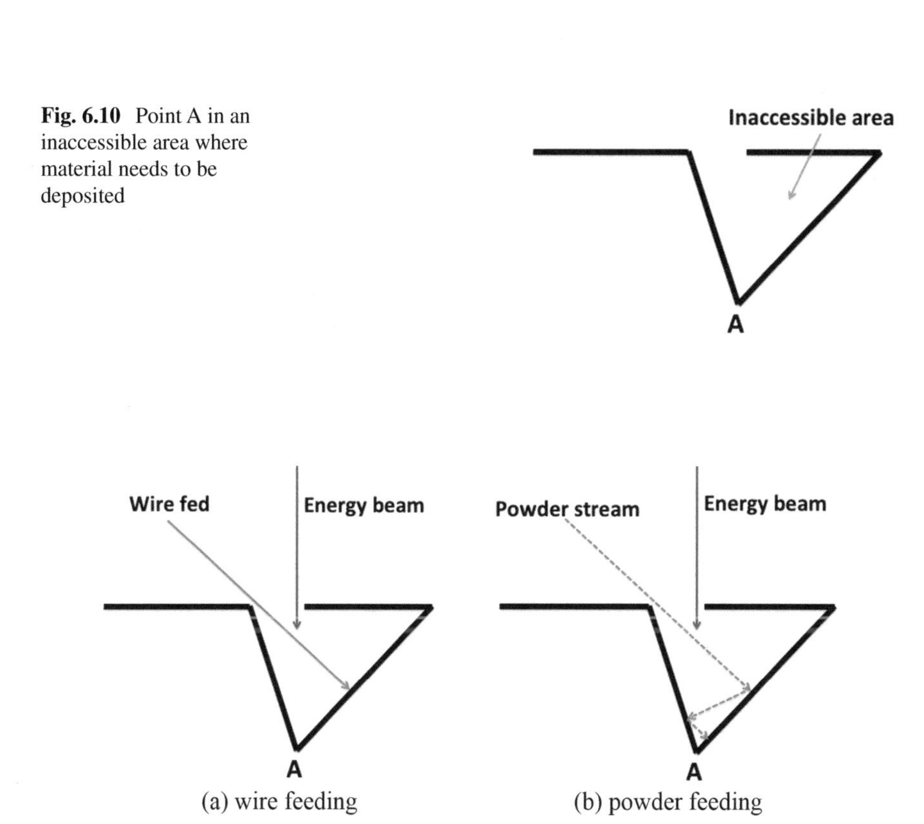

Fig. 6.10 Point A in an inaccessible area where material needs to be deposited

Fig. 6.11 Material deposition at point A of an inaccessible area: (**a**) wire feeding, (**b**) powder feeding

References

Cunningham CR, Flynn JM, Shokrani A et al (2018) Invited review article: strategies and processes for high quality wire arc additive manufacturing. Addit Manuf 22:672–686

Eisenbarth D, Esteves PMB, Wirth F, Konrad Wegener K (2019) Spatial powder flow measurement and efficiency prediction for laser direct metal deposition. Surf Coat Technol 362:397–408

Feng Y, Zhan B, He J, Wang K (2018) The double-wire feed and plasma arc additive manufacturing process for deposition in Cr-Ni stainless steel. J Mater Proc Technol 259:206–215

Geldart D (1973) Types of gas fluidization. Powder Technol 7:285–292

Haley JC, Schoenung JM, Lavernia EJ (2019) Modelling particle impact on the melt pool and wettability effects in laser directed energy deposition additive manufacturing. Mater Sci Eng A 761:138052

Hassen AA, Noakes M, Nandwana P et al (2020) Scaling up metal additive manufacturing process to fabricate molds for composite manufacturing. Addit Manuf 32:101093

Kirka M, Bansal R, Das S (2009) Recent progress on scanning laser epitaxy: a new technique for growing single crystal superalloys. In: SFF proceedings, pp 799–806

Li L, Huang Y (2018) Interaction of laser beam, powder stream and molten pool in laser deposition processing with coaxial nozzle. J Phys Conf Ser 1063:012078

Masood SH (2014) Advances in fused deposition modeling. In: Comprehensive materials processing, vol 10, pp 69–91

McLaskey GC, Glaser SD (2010) Hertzian impact: experimental study of the force pulse and resulting stress waves. J Accoust Soc Am 128(3):1087–1096

Meacock C, Vilar R (2008) Laser powder microdeposition of CP2 titanium. Mater Des 29:353–361

Oliveira UD, Ocelík V, De Hosson JTM (2005) Analysis of coaxial laser cladding processing conditions. Surf Coat Technol 197(2–3):127–136

Pinkerton AJ (2016) Lasers in additive manufacturing. Opt Laser Technol 78(A):25–32

Pirch N, Linnenbrink S, Gasser A, Schleifenbaum H (2019) Laser-aided directed energy deposition of metal powder along edges. Int J Heat Mass Transf 143:118464

Schmidt M, Merklein M, Bourell D et al (2017) Laser based additive manufacturing in industry and academia. CIRP Ann 66(2):561–583

Sergachev DV, Kovalev OB, Grachev GN et al (2020) Diagnostics of powder particle parameters under laser radiation in direct material deposition. Opt Laser Technol 121:105842

Tabernero I, Paskual A, Alvarez P, Suarez A (2018) Study on arc welding processes for high deposition rate additive manufacturing. Proc CIRP 68:358–362

Taminger KMB, Hafley RA (2013) Electron beam freeform fabrication: a rapid metal deposition process. In: Proceedings of 3rd annual automotive composites conference, Troy, MI

Tarasov SY, Filippov AV, Shamarin NN et al (2019) Microstructural evolution and chemical corrosion of electron beam wire-feed additively manufactured AISI 304 stainless steel. J Alloys Compd 803:364–370

Vilar R (1999) Laser cladding. J Laser Appl 11(2):64–79

Vilar R (2014) Laser powder deposition. In: Comprehensive materials processing, vol 10. Elsevier Ltd, Amsterdam, pp 163–216

Watson JK, Taminger KMB, Hafley RA, Petersen DD (2002) Development of a prototype low voltage electron beam freeform fabrication system. In: SFF proceedings, pp 458–465

Wu J, Zhao P, Wei H et al (2018) Development of powder distribution model of discontinuous coaxial powder stream in laser direct metal deposition. Powder Technol 340:449–458

Yan Z, Liu W, Tang Z et al (2018) Review on thermal analysis in laser-based additive manufacturing. Opt Laser Technol 106:427–441

Yin S, Cavaliere P, Aldwell B et al (2018) Cold spray additive manufacturing and repair: fundamentals and applications. Addit Manuf 21:628–650

Yu HZ, Jones ME, Brady GW et al (2018) Non-beam-based metal additive manufacturing enabled by additive friction stir deposition. Scr Mater 153:122–130

Zhang X, Cui W, Hill L, Li W, Liou F (2018) Development of pre-repair machining strategies for laser-aided metallic component remanufacturing. In: SFF symposium proceedings, pp 302–319

Chapter 7
Other Solid Deposition Processes

Abstract It is essential to know arc before knowing arc based additive manufacturing (AM) processes such as gas tungsten arc welding based AM, gas metal arc welding based AM and plasma arc welding based AM. Similarly, it is essential to know cold spray before knowing cold spray based AM. This chapter attempts to describe arc, cold spray and related AM processes. There are friction based processes such as additive friction stir deposition and friction surfacing based AM – these are explained. Extrusion based processes based on both filament and pellet are given.

Keywords Cold spray · Friction stir · Plasma arc · Wire · Fused deposition · Welding · Filament · Pellet

7.1 Introduction

Beam based solid deposition processes have advantages. Beam provides an ability to make fine features and a part having high accuracy. Beam provides an ability to scan – in the absence of such ability the job of scanning needs to be done by moving machine parts. Beam provides an ability to focus on a spot – a deposition with precision can be obtained. Beam provides an ability to defocus – a fast deposition can be obtained by decreasing the scanning time (Steen and Majumder 2010). Beam provides an ability to focus and defocus (Wang et al. 2018) – a structure consisting of various types of depositions leading to various properties at its various features can be obtained. Beam goes far away – a deposition at a remote site of a complex part can be obtained, if repair or refurbishment is an aim. But, beam has certain disadvantages: beam is not absorbed by many materials, beam is expensive. Arc based AM processes come as an alternative as they are not expensive and they can melt higher melting point materials.

Beam based or arc based AM processes depend on melting of the material and have to deal with all problems associated with melting and resolidfication such as porosity, residual stress, crack, anisotropy in properties etc. There are some processes that do not depend on melting such as cold spray or friction based processes.

S. Kumar, *Additive Manufacturing Processes*,
https://doi.org/10.1007/978-3-030-45089-2_7

There are some processes that depend on melting but they do not depend on melting of metals and as such they do not require arc. They also do not require beam, they do not need the power of beam, they do well without utilizing such concentrated source of energy – these heater based AM processes which utilized filament or pellet are given in this chapter along with arc based, cold spray based and friction based processes.

7.2 Friction Based Solid Deposition Process

Using high energy beams to deposit materials has several shortcomings related to the melting of materials by beams. These shortcomings are: (1) loss of nanocrystalline structure of feedstocks – melting and solidification increases the nanocrystalline grain size leading to loss of strength; (2) generation of porosity – high solidification rate does not let trapped gas to emanate, this trapped gas causes circular pores to form; (3) cracks due to mismatch of thermal expansion – when a mixture of two materials (e.g. metal and ceramic) is melted, during solidification both materials will contract in different amounts depending upon their thermal coefficients of expansion causing development of thermal stress which may result cracks; (4) hot cracking – when a mixture of two materials is melted, during solidification high melting point material will solidify earlier leaving low melting point material still in liquid state, cavities or grain boundaries containing such liquid become a source of crack initiation leading to cracks, this brings restriction to the type and number of materials that can be mixed; (5) elements segregation – depending upon the density and melting point of elements, they segregate in a melt pool leading to inhomogeneous distribution of elements; (6) vaporization of materials – in an effort to melt high melting point material of a mixture of high and low melting point materials, if the temperature of the mixture exceeds the vaporization temperature of low melting point materials, the low melting point materials will vaporize causing a change in composition of the melted part from the composition of initial feedstocks, this brings restriction on maximum melting point difference between two materials of a mixture besides narrowing the processing window for material-loss production, a narrow processing window implies limited variation in scan speed resulting in a slow production speed; (7) anisotropic properties – melting is followed by cooling, cooling takes place in a direction, in case of melting of final layer, it is cooled through previous melted layers or substrate causing a grain growth in the build direction, it induces two mechanical properties – one measured in the direction of grain growth and another measured in the direction transverse to the grain growth direction; (8) large grain size – melting layer by layer increases the temperature of the build providing conducive environment for the grain to increase in size, which decreases strength and ductility.

Friction based solid deposition process such as additive friction stir deposition (AFSD) (Yu et al. 2018) and friction surfacing based AM (FSBAM) are AM processes (Dilip et al. 2013), which make parts without raising the temperature more than 90% of the melting point of the material and is therefore free from defects which come along with fusion based or melting based AM processes. The first process (AFSD) is derived from friction stir processing (FSP) while the second process (FSBAM) is akin to FSP minus stirring.

7.2.1 Additive Friction Stir Deposition

In FSP, a rotating pin attached on a shoulder (friction stir tool) enters a solid material and stirs it (Fig. 7.1); pin is an extended tip of a solid cylinder (shoulder); the role of the shoulder is to prevent materials to be extruded out of plane which is caused due to stirring. In AFSD, there is no pin while the shoulder has a thorough hole, the purpose of the hole is to pass the material onto the substrate from the above, the material could be either powder or a rod. Since the rod cannot be handed over to the substrate from the above in the same way as the powder is handed over to the substrate from the above through the hole unless the rod is broken into small pieces; when the rod, often called consumable rod, is abraded against the substrate, the rod loses materials from its surface to the substrate which is deposited in the form of paste (Schultz and Creehan 2014); this is the handing over the material to the substrate through the hole using rod as a medium or a feedstock. In the case of a rod, a rotating and moving rod will deposit material in a line on the substrate while in the case of powders, a moving shoulder will deposit material in a line on the substrate. The hole is small compared to the whole shoulder, the area of the hole facing the substrate is small compared to the solid shoulder area facing the substrate. When a moving shoulder rotates, part of the solid shoulder area encounters already deposited materials – it deforms the material before it could escape and become free; depending on the linear speed of the shoulder, the shoulder will

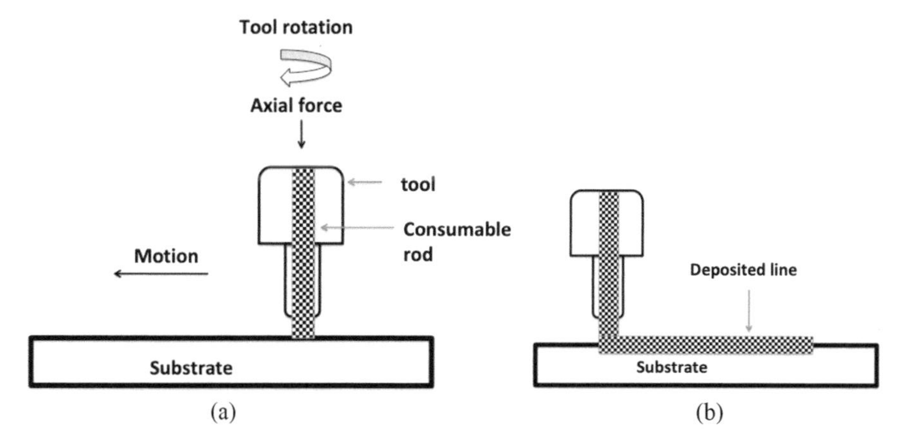

Fig. 7.1 Schematic diagram of AFSD: (**a**) deposition by AFSD, (**b**) deposited line partially in the substrate due to stirring

encounter the same materials time and again and will deform, mix and plasticize–this is termed as stirring.

A deposited material will be free from this cycle of processing if the deposited material is a spherical ceramic powder deposited on a ceramic substrate so that rotating shoulder will be able neither to deform and flatten this powder nor to embed this powder onto the substrate, the powder will then roll over through the gap between the shoulder and the substrate, and will eject. Friction between the shoulder and the substrate increases the temperature, which increases the deformation and stirring of the material. The temperature does not increase more than the melting point of the material so that the benefit realized through solid state processing should not be lost (Yu et al. 2018). High rotation speed of the shoulder may induce melting, which limits the maximum permissible speed, which in turn limits the maximum production speed that can be achieved in AFSD.

AFSD, which is a solid-state processing, in absence of high temperature, there is no driving force for grains to increase in size resulting in small grains – responsible for high strength and ductility (if a number of layers are fabricated, long exposure of initial layers to friction-generated temperature will cause an increase in grain size), stirring and mixing does not let grains to settle resulting in high-angle grain boundaries – responsible for further increase in strength.

This process, like any other friction based AM process, applies axial force for material change (Fig. 7.1a), material joining or transformation; axial force is an essential and unavoidable component of the process, and while the force is good for material transformation, the force is not good for developing or building a structure. For making a structure, the tool which applies force needs to tread over from a thick support to a thin support, to pass through a gap, to make an overbridge, to pass over a bridge, to move in a corner – this may require application of less force to save a feature of the structure (from destruction) or the application of less force just to move. Less force implies no material joining or transformation – this brings to a situation where either material needs to be transformed or the structure needs to be built. This situation can be completely eliminated if no such complex structure will be built – the process will be used to make rectangular, circular, annular, cylindrical or any such blocks. This situation can be partially eliminated if the process is optimized for a range of size of tools so that small blocks could be made with small tools and big blocks could be made with big tools, and the structure is chosen which is a superset of a number of such blocks.

7.2.2 Friction Surfacing Based Additive Manufacturing (FSBAM)

This process is equal to AFSD (using consumable rod) minus stirring. In this process, a rotating consumable rod is heated due to an inter-frictional force between the rod and a substrate (Fig. 7.2), the heating causes the material of the rod to be

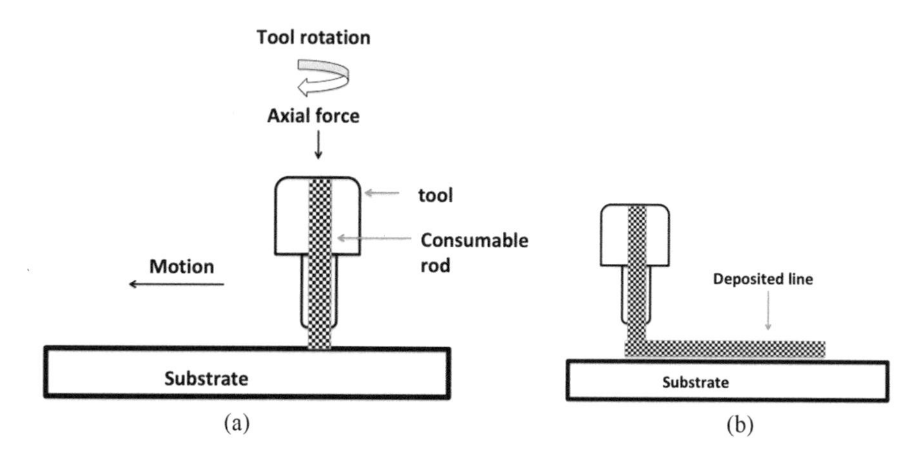

Fig. 7.2 Schematic diagram of FSBAM: (**a**) deposition by FSBAM, (**b**) deposited line over the substrate

plasticized, the material is then smeared on the substrate due to an applied axial pressure on the rod. Instead the rod, the substrate can be heated and plasticized causing the substrate to flow instead, it will not serve the purpose of material deposition by means of a rod. Therefore, the substrate must satisfy following conditions:

1. The substrate should be stronger than the rod so that it is the rod which will plasticize earlier.
2. The substrate should have higher thermal conductivity so that the heat will be dissipated through the substrate while the heat will be accumulated on the rod.
3. The substrate should be relatively cold and the rod be hot – when the rod is hot, it is weaker than the substrate even if both are of same material (e.g. stainless steel, mild steel etc.), it will ensure that same material will be deposited on the same material (of the substrate), it will also ensure that same material (of one layer) will be deposited on the same material (of another layer) to begin multi-layer fabrication, to begin a transition from friction surfacing to AM, to begin limiting the role of the substrate from determining the choice of the deposited material to anchoring and supporting 3D fabrication.

The rod can be heated by frictional heating either by rotating it on some other location of the substrate or on the surface of another plate (Rao et al. 2012), pre-heating is another option (Gandra et al. 2014). The hot rod is then brought to the area of the substrate where it needs to deposit. Cooling the substrate is another method to create a temperature difference between the substrate and the rod. Besides satisfying process requirement, cooling can be used to change grain size (Mishra and Ma 2005).

When a consumable rod begins to deposit, it faces inertia of the substrate, when it exits the substrate, it needs to be detached – this leaves marks on the surface at both entry and exit points, this causes both surface and dimensional inaccuracy. This can be avoided if there is neither an entry point nor an exit point; this is possible if processing is planned in such a way that both points lie outside the boundary of a 3D CAD file and can be trimmed away after the fabrication. The situation similar to arisen by entry and exit points will arise again within the boundary of the 3D CAD file if the rod slows down, takes a turn or deposits on a feature smaller than its diameter. The bonding between a substrate and a layer or between two layers is a metallurgical bonding which is achieved due to the forging (axial pressure of rod) on the plasticized material, thus the properties of the resulting material is similar to a wrought material. The speed of the process depends upon how fast the deposited layer cools down and becomes strong enough for the second layer to be deposited upon.

This process like AFSD does not have means to stir the deposited material, which in turn does not make the material as refined and homogenized as in AFSD. The advantage of the absence of stirring is that small features can be conveniently built without being potentially bent by a stirring action.

7.3 Cold Spray Additive Manufacturing

Cold spray additive manufacturing (CSAM) is a process in which powder particles are accelerated with high speed to deposit on a substrate to make a 3D structure (Fig. 7.3). Cold spray means powder particles are neither completely nor partially melted, and thus in cold spray there is no provision for high temperature sources to melt powders. In thermal spray, partially or fully molten drops are used and thus it requires high temperature sources such as arc, direct current plasma, radio frequency plasma, flame etc. to melt wire, powder or rod to make molten drops. Thus,

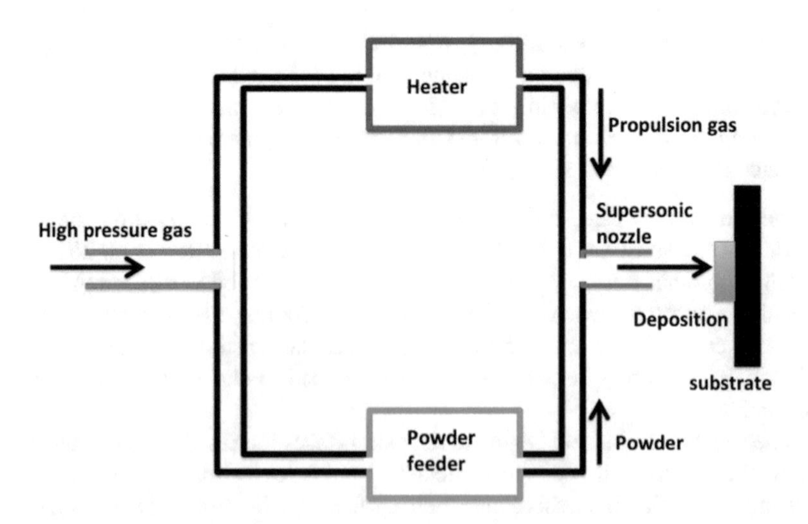

Fig. 7.3 Schematic diagram of CSAM

cold spray is not a thermal spray and thermal spray does not include cold spray. It does not mean that cold spray has nothing to do with high temperature. In cold spray, temperature is increased up to 1000°C but the purpose of high temperature is to increase the speed of the gas. It also does not mean that cold spray has nothing to do with the melting – partial melting of powder may happen when the speed of the powder is high and the substrate is at higher temperature – partial melting will provide metallurgical bonding between powders and the substrate or a deposited layer. Thus, cold spray process is not so cold; it like a thermal spray process is only not hot enough to be able to melt powders before their deposition.

Cold spray relies on the speed of the powder or particles to create bond between particles and the substrate – kinetic energy of a particle provides binding energy while in thermal spray kinetic energy plus thermal energy provides binding energy. If the speed is high, particle will deform, which will break its oxidized layer exposing its clean surface to substrate for metallurgical bonding; particle can also get interlocked in the surface roughness or non-uniformity giving rise to mechanical bond, particles can also be trapped inside the microcrevice of the substrate surface resulting in a bond. If the speed is high but neither the particle nor the substrate is deformable then the particle will reflect back and there will be no bonding. Thus, a combination of high speed and ductility of the material is required. If the substrate is not ductile, then by localized heating its non-ductility can be decreased to some extent. If the particle is not ductile, then by combining it with another ductile material, ductility of resulting composite (ductile plus non-ductile material) can be made workable. Thus, a composite powder made from ductile metal and non-ductile ceramic can be deposited where ductile metal will give rise to bonding while presence of ceramic will impart strength to deposited layers.

A powder which gets deposited without being melted gives certain advantages than a powder which is partially or fully molten during deposition. These advantages are: no phase transition of materials, no residual stress which is caused by solidification, no need for heat management, no grain growth and thus nanostructure of the material can be retained, no possibility of deleterious intermetallic compound formation, no splat, no oxidation in presence of oxidising gas and thus no need for processing chamber and thus no limitation in part size, no thermal expansion mismatch during multi-material deposition etc. – thus, CSAM has advantages over powder fusion based AM.

7.3.1 How Cold Spray Is Generated

If high speed gas is available and some powder is injected on it, then there is cold spray which can be used for AM. High speed gas coming from a pressurized cylinder or coming from some centralized pressurized gas source will fulfil one of the basic conditions of cold spray. The maximum pressure of the gas thus obtained is limited which limits the highest speed obtained therefrom. The speed thus obtained may not be sufficient. If gas is heated on its way, then the pressure and thus the speed obtained can be further increased. The speed thus obtained again may not be sufficient. This brings a question whether there is a method to increase the speed of the gas without using increasingly powerful cylinders and increasingly hotter heaters.

If gas and or gas plus accelerated powders exit through a cylindrical pipe to the substrate, then the speed will remain same because the cylindrical pipe will not be able to increase the speed. If the diameter of the pipe at the end of the cylinder or the exit diameter is constricted or made smaller then the speed will increase if the gas is not compressible. Since there is no gas generated inside the pipe, the gas entering to the pipe must be same as the gas leaving the pipe – the amount of gas entering per unit time must be same as the amount of gas leaving per unit time. The amount of gas that can be accommodated in a pipe depends upon its diameter – if the diameter of the pipe is decreased then it can accommodate less amount of gas and if the flow rate will not increase then the amount of gas entering per unit time will be more than the amount of gas leaving per unit time, and therefore the rate increases. Therefore, by taking a cylindrical pipe having constricted diameter at one end or taking a conical nozzle the gas speed can be increased without taking resort to extra cylinders or heaters. But, the speed thus achieved will still be limited. Moreover, the gas will diverge which will increase the diameter of deposited tracks and will not be suitable for making small features.

If the gas is compressible, then the amount of gas that can be accommodated in a pipe will not only depend on the diameter but also on the compressibility of the gas, and for a compressible gas, same advantage with a conical nozzle may not be obtained. Therefore, the increase in speed of the compressible gas may not be as high as an increase in the speed of an incompressible gas. However, if the increase in speed is sufficient enough to increase the speed of compressible gas more than or equal to the speed of sound then this high speed gas (supersonic gas) increases the compressibility of the gas. Since compressibility increases, the gas no longer follows the basic principle of an incompressible gas – the principle is when the incompressible gas will pass though a convergent nozzle, its speed will increase and when the same gas will pass though a divergent nozzle, its speed will decrease. But, the gas follows the basic principle of a very compressible gas – the principle is when the compressible gas will pass though a convergent nozzle, its speed will decrease and when the same gas will pass though a divergent nozzle, its speed will increase. At such supersonic speed resulting in high compressibility, the speed of the compressible gas further increases when it enters a divergent nozzle or when it exits from a convergent section to a divergent section of a convergent-divergent nozzle.

Thus, a gas coming from a compressed cylinder gets its speed increased in the convergent section of a convergent-divergent nozzle or de laval nozzle because the gas is slightly compressible (behaving like an incompressible gas), and the speed gets further increased in the divergent section of the de laval nozzle because the gas is no longer just slightly compressible but has become highly compressible because of the change in behaviour of gas at such high speed.

Thus, using a right type of nozzle is another method to increase the speed of gas. When particle is injected in this gas, the particle is accelerated at high speed and can be used for cold spray.

7.4 Arc Welding Based Additive Manufacturing

Arc created between two electrodes is used as a heat source in metal welding. This technique of metal welding is extended to create 3D structures, and the resulting group of processes is arc welding based additive manufacturing. The group consists of AM processes based on gas tungsten arc welding (GTAW) (Fig. 7.4), plasma welding (plasma non-transferred arc, plasma transferred arc) (Fig. 7.5) and gas metal arc welding (GMAW) (Fig. 7.6). When one or all of these arc based processes are applied to fabricate from wire feedstock, the process is named as wire arc additive manufacturing (WAAM) (Tabernero et al. 2018; Cunningham et al. 2018). These processes are marked by inexpensive heat sources, inexpensive equipment and their potential to make large components.

7.4.1 What Is Arc

When voltage is applied between two electrodes, the air molecules or atoms present between tips of the electrodes get polarized – it means positive charge and negative charge of an atom gets separated from each other creating positive and negative poles within an atom. If voltage is increased further then these positive and negative poles are no longer confined within the boundary of the atom and gets detached from each other – this is called ionization when atoms and molecules between the tips of electrodes get converted into positive and negative charges. Thus, air between tips consists of charges – these charges are collectively called plasma. Negative

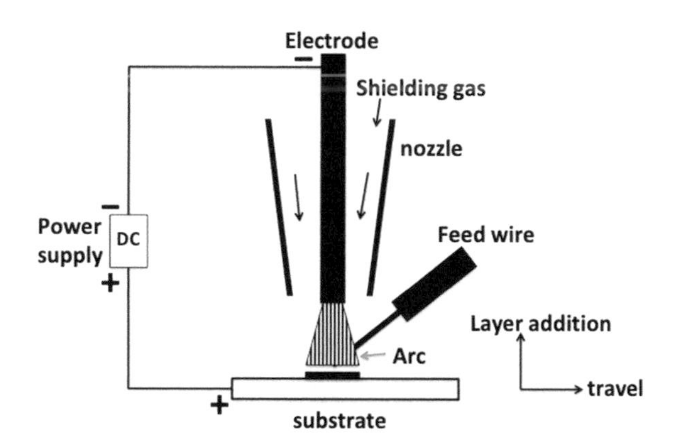

Fig. 7.4 Schematic diagram of GTAW based AM

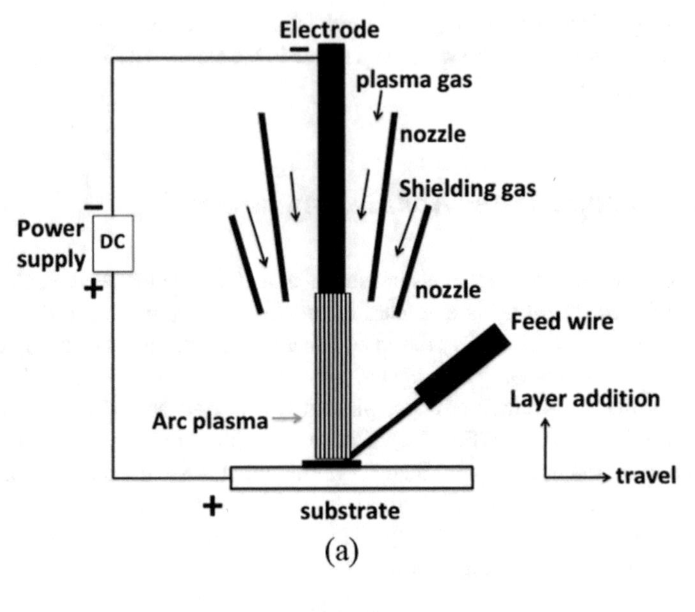

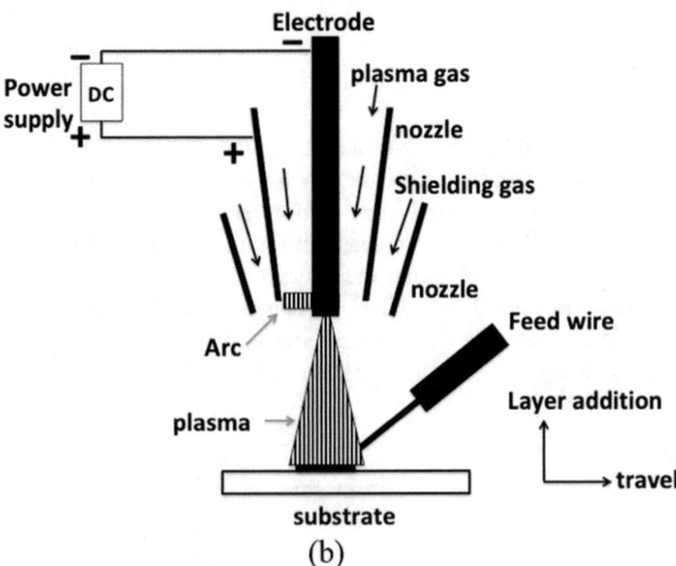

Fig. 7.5 Schematic diagram of plasma welding based AM: (**a**) transferred arc based AM, (**b**) non-transferred arc based AM

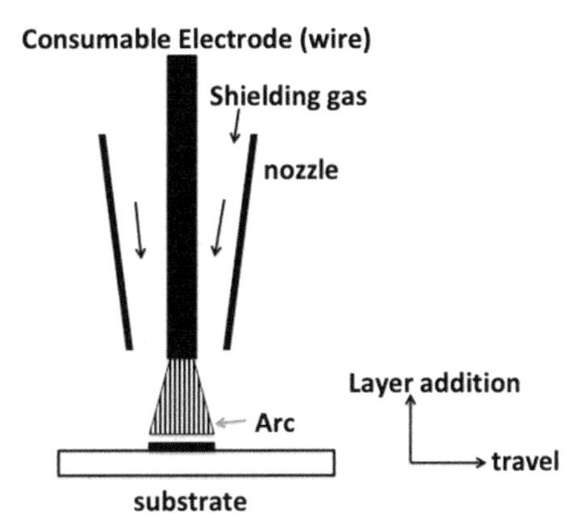

Fig. 7.6 Schematic diagram of GMAW based AM

charges will move towards positive electrode and positive charges will move towards negative electrode – this movement gives rise to an arc and thus such motion in charges is called an arc. Therefore, arc is a visible and shining line between electrodes. Movement of charges faces obstacles due to the presence of other particles and molecules in the air, and in the course of overcoming these obstacles, friction occurs between moving charges and particles which gives rise to heat – thus, creation of arc entails creation of a heat source. Higher voltage means more ionization, stronger attraction and more intense arc giving more heat. Higher current means more injection of electrons in plasma, thus more ionization and stronger arc.

Ionization not only depends on voltage and current but on the type of gas present between the electrodes as well. Gas having lower ionization potential will be ionized at lower voltage than the gas having higher ionization potential. Thus, argon gas (ionization potential 15.7 eV) is easier to be ionized and used at lower voltage than helium gas (ionization potential 24.5 eV). Gas not only decides amount of ions in an arc, it also decides whereabout of heat in an arc. If the gas has high thermal conductivity, such as helium gas (thermal conductivity 0.151 W/mK), then the heat will no longer be confined within the arc line, heat will spread out making the influence of arc wider – it will result in wider welding bead. If the gas has low thermal conductivity such as argon gas (thermal conductivity 0.018 W/mK), then heat will not spread which will result in deeper welding rather than wider welding. Deeper welding requires an arc containing high heat, which might not be possible by just having one favourable physical property, that is low thermal conductivity of the gas. If the gas in an arc furnishes extra heat, then it will serve the purpose. Oxygen gas during ionization splits into oxygen atoms and furnishes extra heat (dissociation energy) – thus, presence of oxygen will contribute to the intensity of an arc. If small amount of oxygen gas is mixed with argon gas, then the intensity of the arc will increase. Though, oxygen gas (thermal conductivity 0.027 W/mK) has slightly higher thermal conductivity than that of argon gas, it will not offset gain in heat intensity obtained due to the dissociation of oxygen gas.

Heat generated in arc affects electrodes – these can be melted and can be used as a source of materials for AM, or the electrode can be degraded when the electrode is not supposed to be melted but is supposed to act just as an electric terminal to create arc. In gas tungsten arc welding based AM, tungsten electrode is supposed to act as an electrical terminal to create arc. Cooling the electrode is one of the ways to mitigate heat-induced degradation. Cooling also happens when gas flows by the electrodes, this is the same gas which is a precursor for the arc. If the gas has high thermal conductivity, then the heat transfer will be more and the electrode will cool more – therefore, helium gas which has higher thermal conductivity cools electrodes more than the argon gas does.

When direct current is flowing and tungsten electrode is maintained as positive electrode, degradation of the electrode is faster because positive tungsten ions from it are pulled by negative electrodes. It gives rise to the loss of tungsten atom from the electrode resulting in depletion of electrodes. Using alternating current instead of direct current helps avoid this type of degradation. When tungsten electrode is maintained as negative electrode then it does not have any such degradation through loosing positive ions; besides, heat is generated more at the positive electrode rather than at negative electrode and therefore for the same setting of current and voltage it is the positive electrodes which becomes hotter and degrades more.

Negative ions which are generally electrons and smaller, and bigger positive ions are constituents of an arc. When voltage is applied across electrodes, negative ions being smaller move faster towards positive electrode while positive ions being bigger move slower towards negative electrode. Faster electrons have higher impact on the positive electrode making it hotter. Thus, a tungsten electrode which is maintained as a negative electrode is cooler and safer; its efficiency is further improved by alloying it with a small amount of thoria. When voltage is applied across electrodes, it is the negative electrode from which electrons enter into the arc, the transfer of electrons from electrode to the arc depends also upon the crystal structure of the electrode. When tungsten is mixed with thoria or other materials which decrease the work function of the tungsten then transfer of electrons becomes easier and the efficiency of the electrode increases. Addition of 2 wt. % thoria decreases the work function of tungsten from 4.5 eV to 2.6 eV.

7.4.2 Gas Tungsten Arc Welding (GTAW) Based AM

This process as the name implies uses a tungsten electrode to create arc, while the prefix gas implies that shielding gas is used to prevent oxidation of materials which are going to be melted during welding or AM. Shielding gas continuously flows around arc so that the shielding gas shields atmospheric gas or air to come near to the arc and thus prevents reaction between atmospheric gas and arc-metal interaction zones (Fig. 7.4). Shielding gas is thus generally an inert gas such as argon or helium but can also be a mixture of inert and non-inert gases. Shielding gas is the same gas from which an arc is formed and thus it serves the dual purpose of shielding and arc-creating. Since the type of arc depends upon the type of gas, selection of a shielding gas does not entirely depend upon the sole purpose of shielding but depends upon the type of arc needed.

Tungsten electrode is made cathode and a work-piece or substrate is made anode to strike an arc when direct current is used. Since 70% of heat is generated at the anode, this type of polarity will help over-heat the workpiece and under-heat the cathode – since the aim is to melt workpiece (or the material on the workpiece) rather than the tungsten electrode, this type of polarity will help conserve energy by channelizing the heat near the workpiece. For creating a 3D structure, more than a combination of tungsten electrode, workpiece and arc is required – materials in the form of powder or wire are required. These materials are fed to be melted either directly by arc or indirectly by meltpool created by the arc. Since the stability of the arc remains largely undisturbed by its effect on melting, it gives an advantage for controlling the deposition. Most of the research in AM is yet done by feeding wire either coaxially or sidewise.

7.4.2.1 Plasma Welding Based AM

The arc created by tungsten electrode in GTAW exists near the electrode and thus the operational ability of the arc is limited by its physical proximity to the electrode. If the arc were created far away from the electrode, then it could have more freedom – the effect of melting due to arc could have been planned without worrying about its consequence on the tungsten electrode, the feeding of materials could not have been restricted due to constricted space between the electrode and the workpiece, the size of the melt pool could not have been limited due to the limited variation in arc dimension. These limitations are in sharp contrast to the freedom provided by a laser beam – the effect of a laser beam spot on a workpiece does not affect a laser source or laser device responsible for generating the laser beam.

The limitation imposed by such arc can be overcome by creating better heat sources utilizing such arcs. There are some methods to make a better heat source using arc created by tungsten electrode. Since arc is hot, if a gas will pass through the arc then the gas will become hot and such hot gas can be used as a heat source. Thus, one of the methods to overcome such limitations is to create a hot gas as a heat source with the help of an arc. In this method, a tungsten rod is fitted inside a hollow nozzle while the rod usually acts as a cathode the nozzle acts as an anode. When a gas is flown through the nozzle from its top, the gas provides a medium to create an arc between the rod and the nozzle. Besides, the gas coming from the nozzle is affected by the arc that is already created – the gas is no longer the same gas which entered into the nozzle, the gas is hot and ionized. Since the gas is ionized and has become a mixture of positive and negative ions, it is termed as plasma. This plasma can reach away from the location of the cathode, it is no longer confined near the cathode as the arc is, and therefore the effect of the plasma is no longer confined to the workpiece kept near the cathode but can be realized at a longer distance. The reach of the plasma depends upon the speed of the gas entering through the nozzle, it also depends upon the type of the nozzle and the hotness of the arc.

This new heat source is due to thus generated plasma; this new heat source is due to the arc struck between a nozzle and a cathode – in order to produce this new heat source, the location of the arc does not need to be changed. It means that arc remains maintained between the nozzle and the cathode before the flow of plasma through the nozzle to the workpiece and after the flow of plasma – if the very arc remains maintained then the arc is not transferred anywhere. Therefore, this new heat source is called plasma non-transferred arc (Fig. 7.5b).

It brings a question – what if an arc is no longer maintained between a nozzle and a cathode – and the arc is transferred to the workpiece. Then, the arc striking the workpiece and heating the workpiece is no longer same as plasma striking the workpiece and heating the workpiece – this is a new method of heating the workpiece and there could be one more type of heat source possible by utilizing the arc between the nozzle and the cathode. In this method, an arc is firstly created between a nozzle and a cathode, since this arc does not serve or intend to serve either directly or indirectly as a heat source, a less hot arc or low-current arc is sufficient which is started between the nozzle and the cathode. When the gas passes through it and plasma strikes between the cathode and the workpiece, it is easier to create another arc between the cathode and the workpiece. Thus, when polarity of the workpiece is changed from neutral to positive with respect to the tungsten rod, another arc is started. The initial arc created serves its purpose when this new arc is created and the initial arc is then extinguished; the initial arc is named as pilot arc while the heat source due to this new arc is called plasma transferred arc (Fig. 7.5a). Pilot arc is also started by a high-frequency generator and in that case the polarity of the nozzle and the tungsten electrode periodically changes.

Plasma transferred arc needs to pass through a constricted nozzle orifice and is therefore thin resulting in the concentration of heat energy. Since plasma gas also passes through the orifice, the gas is converted into plasma due to its interaction with the arc which results in a heat source consisting of both arc and plasma. Therefore, this heat source has higher temperature (almost two times) than the arc in GTAW does have. Besides in plasma transferred or non-transferred arc, shielding gas is separately used from plasma gas which provides an advantage that the striking of arc is not affected by the shielding gas as much as it is affected in GTAW.

7.4.3 Gas Metal Arc Welding (GMAW) Based AM

When an arc is created between two electrodes and one electrode is melted and ready to be deposited then the need for creating provisions for feeding materials is eliminated or minimized. In GMAW, out of two electrodes, one electrode is consumable and other electrode is workpiece or substrate (Fig. 7.6). If one electrode is consumable, it means it is feedstock, or feedstock to this process is supplied in the form of an electrode. Since the form of a consumable electrode is usually rod or wire, the feedstock to this process is usually rod or wire rather than powder. Since the consumable electrode is same as a feedstock wire, the process gives an advantage to efficient transfer of heat from the arc to the wire in comparison to a process where electrode and wire are not the same – this results in a fast deposition of materials. If having consumable electrode means advantage to the process, then having consumable electrode means also disadvantage to the process. If an electrode is getting consumed, then arc length will no longer be same, and if the arc length will change, then it will disturb the process by changing either voltage or current which will require further adjustment of voltage, current or feed rate – all these problems of adjustment are not required in a process where electrode is not consumed.

The process makes a 3D part by transferring molten parts of an electrode from an electrode to a workpiece (Fig. 7.6). Besides other factors, the transfer of molten parts from the electrode depends mainly upon how much molten the molten part is. If it is little bit molten, mainly at the condition of low current, then this will not be detached from the electrode on its own and therefore no deposition will occur. If it is little bit more molten, mainly at the medium-current condition, then this will be detached from the electrode on its own and therefore deposition will occur. In order to be detached, gravitational force needs to overcome surface tension. After detachment, droplets which are approximately of the size of the diameter of the electrode will be detached. If it is well molten, mainly at the high-current condition, then it will not only be detached from the electrode on its own in the form of small droplets. Because, at this condition, there will be high heat energy which will decrease the surface tension force and therefore small gravitation force due to a small droplet will be able to overcome the surface tension force. With an increase in current, heat energy increases which increases hotness of molten material allowing to overcome surface tension forces fragmented into droplets. Instead of increasing current, this increase in heat can also be imparted by other methods such as use of oxidizing gas, and a similar trend can be obtained. These three conditions of current broadly show three types of material transfer process: short-circuit transfer related to low current, globular transfer related to medium current and spray transfer related to high current.

In short-circuit transfer, since material is not detached on its own, it needs to be detached by initiating a short circuit between electrode and work-piece. When material is not getting detached from the tip of the electrode or wire then the wire is continuously fed which continuously decreases the arc length between the wire and the workpiece until the arc length or gap becomes zero, this is short-circuit. A constant current then passes from the wire to the workpiece which gives rise to resistive heating at the interface between the wire and the workpiece; this heat melts the tip of the wire. Molten material detaches from the tip of the wire by overcoming surface tension at the tip; this again creates an arc between the tip of the wire and the deposited material which brings an end to the short-circuit. Thus, in this type of material transfer, a continuous cycle of creation of short-circuit and end of short-circuit gives rise to continuous deposition of materials and fabrication of a part. In this type, uncontrolled resistive heating produces spatter which decreases the precision of deposition. In order to avoid resistive heating and reduce noise and spatter, a variation of this type of metal transfer is developed; this variation is called cold metal transfer (CMT), the prefix 'cold' of this process denotes the absence of (resistive) heating. In CMT, current is set zero and the wire is retracted the moment semi-molten tip of the wire touches the workpiece, and thus resistive heating is avoided. In absence of resistive heating, the tip of the wire does not get heat energy from the resistive heating and needs to melt without the resistive heating. Thus, in CMT, low current is not low enough not to melt the tip of the wire, and the low current is not high enough to detach from the tip of the wire on its own. CMT is able to furnish 3D steel structures with high deposition rate and reproducibility (Ali et al. 2019).

7.5 Extrusion Based Additive Manufacturing

Extrusion in extrusion based AM implies extrusion of polymer based materials and thus it is different from the meaning of extrusion in general manufacturing where it can be used even for metals and alloys. Extrusion of polymers in general manufacturing may also mean softening and drawing of polymers but extrusion in AM is not related to drawing. Extrusion in AM means applying a force on a polymer based material so that the material will flow through a nozzle (extruder head or printer head) in a continuous shaped form. This shaping of the material after it is sent through a nozzle and before it reaches a substrate gives the process its name. This shaping of the material distinguishes this AM process (or this family of AM processes) from other AM processes (Chi et al. 2017) which rely on nozzles to send liquid through. This does not mean if there are two types of shaping coming from two types of nozzles, then it will bring substantial difference between AM parts made therefrom. This only conveys that absence of such shaping means absence of such processes. This also conveys that the role of a nozzle is not only to determine the quantity of materials sent through or to ensure the flow of materials or the focus of the materials but also to shape the materials. This does not mean that the role of the nozzle to shape the materials is more important than other roles of the nozzle. This only means that this role (of shaping) of the nozzle gives a family of AM processes while other roles of the nozzle do not give such family of processes. This does not mean that the shaping is solely decided by a nozzle and roles played by types of materials, temperature, environmental conditions, applied force have no influence on shaping. This only means that the role of a nozzle has more ability to separate this family of AM processes from other AM processes. This family of AM processes is fused deposition modeling (FDM) (Masood 2014) also known as fused filament fabrication (Brenken et al. 2018), fused pellet modeling (FPM) (Wang et al. 2016) or fused layer modeling (Kumar et al. 2018) also called big area additive manufacturing (BAAM) when making big parts (Roschli et al. 2019), powder melt extrusion (PME) (Boyle et al. 2019), composite extrusion modeling (CEM) Lieberwirth et al. 2017) etc.

Extrusion in AM is broadly related to the viscosity of polymer based materials at a given temperature of extrusion – if viscosity is high, material will not flow out of the nozzle even after the application of force; if viscosity is low, material will flow out of the nozzle due to gravity (or material will fall) and it will not be possible to control the flow of material by application of force, and thus it will not be possible to shape the material through the nozzle, and therefore the resulting process will not be an extrusion based AM. Therefore, an optimum viscosity is required which will enable to produce a continuous shaped material. What if a nozzle is very fine so that even for a low viscous polymer, application of force is required to push the material through the nozzle – this will come under extrusion or non-extrusion and the resulting AM will be extrusion based AM or non-extrusion based AM. It will depend upon the consequence of application of force. If the consequence is a droplet or a stream of droplets, then there is no shaping because droplets are formed due to gravity; in this material, it is the gravitational force that is determining the shape of the material, and the application of force through the nozzle has no role in shaping; besides, a stream of droplets makes a discontinuous body and not a continuous integrated body, thus, this is non-extrusion and resulting AM is not an extrusion based AM. If the consequence of application of force is a continuous shaped material which does not break due to gravity and is able to retain its shape then this is extrusion and resulting AM is an extrusion based AM.

7.5.1 Feedstock Type in Extrusion Based AM

Filaments, pellets and powders are used as feedstock in extrusion based AM. Filament based process is called fused deposition modeling (FDM) or fused filament fabrication (Fig. 7.7a). Pellet based process is called fused pellet modeling (FPM) or fused

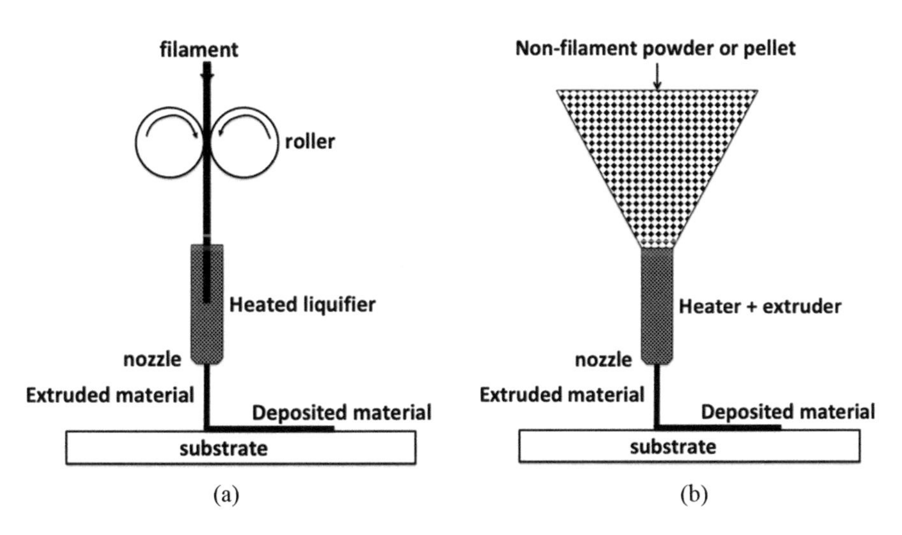

Fig. 7.7 Schematic diagram of extrusion based AM: (a) filament type extrusion based AM, (b) non-filament type extrusion based AM

layer modeling. Powder based process is called powder melt extrusion (PME) and composite extrusion modeling (CEM) (Fig. 7.7b).

Filaments are widely used feedstock and almost all commercial systems are made to use filaments while few systems use pellets. The advantage of filaments is that filaments not only work as a feedstock but also work as a machine part of an FDM system – a filament is used to push the molten material through a nozzle, or a moving filament acts as a piston to push molten filaments. Thus, filament brings simplicity in extrusion based AM system development. This is one of the reasons why filaments based systems are in abundance. But, what is the reason for being an advantage of a filament is also the reason for being a disadvantage of a filament. If a filament is to act as a piston, then it must possess more properties than a filament as a feedstock needs to possess – this is the disadvantage, this requires extra effort in its development, this places extra conditions on a material to be accepted as a filament. A filament should have enough rigidity that it will itself not buckle when it pushes the molten material but the filament should not have so much rigidity that it will not bend when it is driven from its source. A filament should not be so much weak that it will break when it pushes the material and the filament should not be so much strong that it will increase its melting point so much that it does not melt. Thus, a filament needs to satisfy conflicting property requirements which exclude many materials to be converted into filaments and thus exclude many materials to be converted into 3D parts though filament based processes.

If there is an extrusion based AM process which uses feedstock other than filaments, then the process will not be deprived of the advantage gained from extrusion as well as the process will be free from the disadvantages of using filaments. If feedstocks other than filaments such as pellets and powders are used, then there will be difference in how they are brought to the print head for melting and extrusion – they will need different setup for storing the feedstock and an extra screw or piston for extruding; but there will be no difference after they will be melted and extruded. Using pellets or powders provide opportunities to use various types of polymers and their mixtures and make 3D parts without fulfilling the requirements of a filament. This has facilitated to use elastomeric materials, ceramic and metallic materials in extrusion based process. A metallic product can also be formed by using a composite feedstock made from metal and polymer and removing polymer materials from an extruded 3D part in downstream processes.

7.6 Comparison Between Friction Based SDP and Fusion Based SDP

- *Small features*: In friction based process, fabrication of a small feature is possible if it is able to withstand tool pressure during fabrication while in fusion based process there is no tool pressure and the smallest features depend upon the minimum diameter of a high energy beam.

- *Material properties*: Fusion based process gives material properties similar to a cast material which can differ due to direction of solidification, rate of solidification and thermal gradients while friction based process gives material properties similar to a wrought material (Schultz and Creehan 2014).
- *Processing an inaccessible area*: A high energy beam can reach to farthest inaccessible area in comparison to a friction tool, therefore fusion based process is better than friction based process when a modification needs to be done or a small feature needs to be added in an inaccessible area.
- *Tool wear*: In friction based process, tool gets worn out which needs to be replaced, there is no such tool wear in fusion based process.
- *Flexibility*: Fusion based process can make a part having features of various sizes, while friction based process does not have such flexibility due to fixed size of a tool.
- *Retaining the original crystal structure*: If the aim is to retain the original crystal structure of feedstock then friction based process because of being solid state process and low-temperature process is a better alternative than the fusion based process. Because melting in fusion based process will change crystal structures and after solidification, the structure will not be the same.

References

Ali Y, Henckell P, Hildebrand J et al (2019) Wire arc additive manufacturing of hot work tool steel with CMT process. J Mater Process Technol 269:109–116

Boyle BM, Xiong PT, Mensch TE et al (2019) 3D printing using powder melt extrusion. Addit Manuf 29:100811

Brenken B, Barocio E, Favaloro A et al (2018) Fused filament fabrication of fiber-reinforced polymers: a review. Addit Manuf 21:1–16

Chi B, Jiao Z, Yang W (2017) Design and experimental study on the freeform fabrication with polymer melt deposition. Rapid Prototyp J 23(3):633–641

Cunningham CR, Flynn JM, Shokrani A et al (2018) Invited review article: strategies and processes for high quality wire arc additive manufacturing. Addit Manuf 22:672–686

Dilip JJS, Babu S, Rajan SV et al (2013) Use of friction surfacing for additive manufacturing. Mater Manuf Proc 28:1–6

Gandra J, Krohn H, Miranda RM et al (2014) Friction surfacing- a review. J Mater Proc Technol 214(5):1062–1093

Kumar N, Jain PK, Tandon P, Pandey PM (2018) Investigation on the effects of process parameters in CNC assisted pellet based fused layer modeling process. J Manuf Proc 35:428–436

Lieberwirth C, Harder A, Seitz H (2017) Extrusion based additive manufacturing. J Mech Eng Automat 7:79–83

Masood SH (2014) Advances in fused deposition modeling. Compr Mater Process 10:69–91

Mishra RS, Ma ZY (2005) Friction stir welding and processing. Mater Sci Eng Rep 50(1–2):1–78

Rao KP, Sankar A, Rafi HK (2012) Friction surfacing on nonferrous substrate: a feasibility study. Int J Adv Manuf Technol 65(5–8):755–762

Roschli A, Gaul KT, Boulger AM et al (2019) Designing for big area additive manufacturing. Addit Manuf 25:275–285

Schultz JP, Creehan KD (2014) Friction stir fabrication. US Patent (US 8893954 B2)

Steen WM, Majumder J (2010) Laser material processing. Springer-Verlag London Limited, London

Tabernero I, Paskual A, Alvarez P, Suarez A (2018) Study on arc welding processes for high deposition rate additive manufacturing. Proc CIRP 68:358–362

Wang Z, Liu R, Sparks T, Liou F (2016) Large scale deposition system by an industrial robot (I): design of fused pellet modeling system and extrusion process analysis. 3D Print Addit Manuf 3(1):39–47

Wang L, Zhu G, Shi T et al (2018) Laser direct metal deposition process of thin-walled parts using variable spot by inside-beam powder feeding. Rapid Prototyp J 24(1):18–27

Yu HZ, Jones ME, Brady GW et al (2018) Non-beam-based metal additive manufacturing enabled by additive friction stir deposition. Scr Mater 153:122–130

Chapter 8
Liquid Based Additive Layer Manufacturing

Abstract Liquid based AM processes are varied: in one extreme very big parts can be made by photopolymerization and in the other extreme thinner lines are deposited for electronics applications; besides, this is the liquid which has started non-layer based AM processes. This chapter has brought all processes together, though it deals only with layer based processes. While photopolymer bed and liquid deposition process are briefly mentioned, water based process and slurry based process are dealt with in somewhat detail. This chapter has reasoned why the name stereolithography is illogical, and thus photopolymer bed process described in Chap. 2 can instead be a better name. Four-dimensional printing utilizes liquid based AM processes and therefore its relation with liquid based AM processes is mentioned.

Keywords Ink jet · Stereolithography · Photopolymerization · Rapid freeze · Cryogenic prototyping · Slurry

8.1 Introduction

A liquid is defined as a substance that has a constant volume and that moves freely. Hence, a liquid has no independent shape and when it is poured in a container, it takes the shape of the container. In additive manufacturing (AM), there are an abundance of liquids that are used such as polymer, photopolymer, ink, water, gel, slurry, molten metal (Jayabal et al. 2018) etc. These liquids are clearly different from other materials used in AM such as powder, wire, filament, gas, rod etc. and therefore liquid (or a state of matter) can be used to distinguish one AM from others. As liquid is defined as a substance which takes the shape of a container when it is poured into the container – this brings a question how liquid will take the shape of a container – it will take the shape on its own without the application of an external force or it will take the shape when an external force is applied on it. If an external force is not applied then the liquid will be of low viscosity such as water or low-viscous photopolymer and if an external force is applied then the liquid will be of high viscosity such as slurry or a photopolymer having high percentage of ceramics. If an external

S. Kumar, *Additive Manufacturing Processes*,
https://doi.org/10.1007/978-3-030-45089-2_8

force is applied then the machine will have provision to apply such an external force and consequently the machine will no longer be same. In both cases, machines will be different; the name of the process may be different; parameters to process will be different; application may be different. In case of a low-viscous photopolymer, the name of the process is stereolithography (SL) while in case of a high-viscous photopolymer based material the name of the process is lithography based ceramic manufacturing (LCM) (Harrer et al. 2017). In case of low-viscous liquid such as water, the name of the process is rapid freeze prototyping (RFP) (Bryant et al. 2003) while in case of high-viscous liquid such as slurry, the name of the process is 3D gel printing (3DGP) (Ren et al. 2016).

Thus, there are a number of processes which have different names, may not have same machines, may not work with the same materials, may not have the same applications, may not have same binding methods, but there is a common thread among these processes which can relate them. The common thread is that they all use materials which can come under the realm of liquid. Recognizing this thread provides a means to bring together all processes under a common name, which is liquid based AM process.

Liquid based AM process can be divided into two categories depending upon the use or non-use of layer to achieve AM. These two categories are liquid based additive layer manufacturing (ALM) and liquid based additive non-layer manufacturing (ANLM) as shown in Fig. 8.1. Processes such as two-photon polymerization (Zhou et al. 2015), CNC accumulation (Chen et al. 2011) and continuous liquid interface production (Janusziewicz et al. 2016) come under liquid based ANLM and are dealt in Chap. 10 while liquid based ALM is dealt in this chapter.

Liquid based ALM is divided into two major categories: bed type and deposition type named as liquid bed process and liquid deposition process respectively. Liquid bed type can be further divided into two types such as photopolymer bed type and slurry bed type while liquid deposition type can be divided into many types such as polymer, photopolymer, ink, water, metal and slurry.

8.2 Photopolymer Bed Process

Photopolymer bed process (PPBP) is better known as stereolithography (Jacobs 1992) or by the name of its variants such as microstereolithography (Bertsch et al. 1999), digital light processing (Santoliquido et al. 2019), large area maskless photopolymerization (Rudraraju and Das 2009) etc.

In PPBP, photo sensitive polymers are used; these polymers are irradiated with light (generally UV laser or other lasers); photons of light interact with electrons and molecules of polymers. These interactions create either free radicals by breaking bonds or excited ions (cations) by removing electrons. These free radicals or cations bond with other polymer molecules which enable to bond with other polymers resulting in an increase in chain length, viscosity, gelling, solidification and increase in molecular weight. Thus, free radicals or cations trigger polymer chain

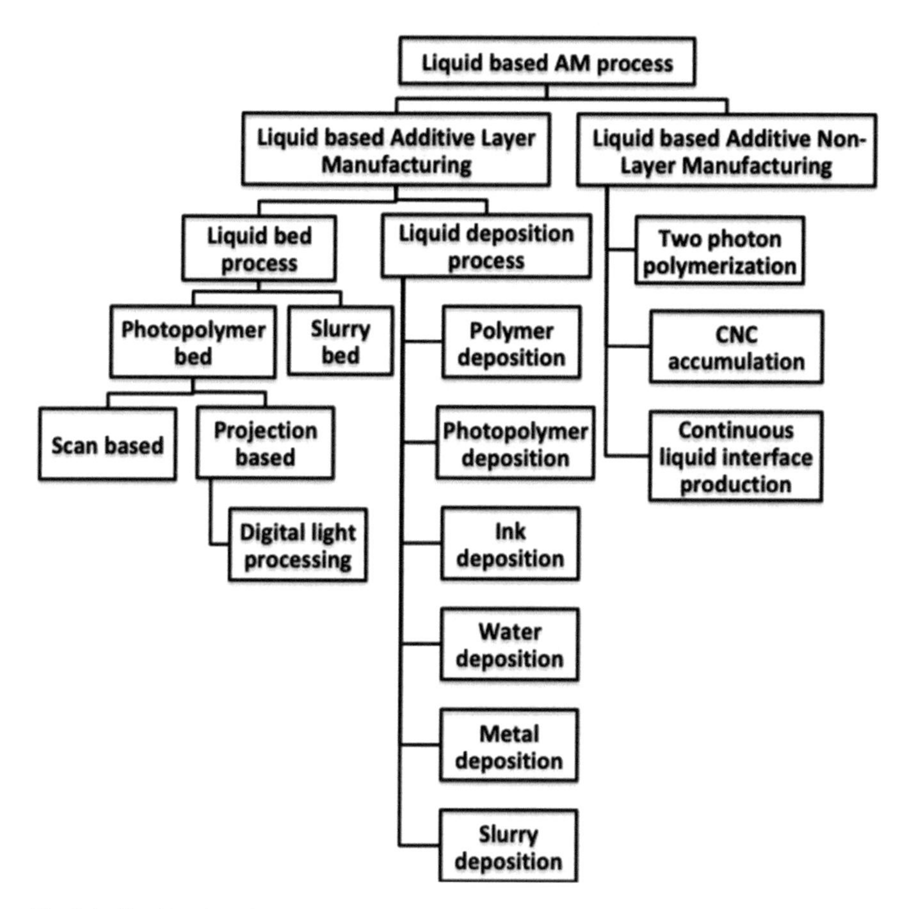

Fig. 8.1 Classification of liquid based process

formation also called polymerization. Since few photons are required to trigger such chain formation and solidification in comparison to the number of photons required to break metallic bonds and melt them, therefore, laser power required in PPBP is of few mW while the laser power required in other AM is of 100 watts.

Since photopolymer is liquid, the process can work in many orientations; this could not be possible if photopolymer were solid particles. In majority of cases, the process works in a general orientation (Fig. 8.2a) where the laser beam is irradiated on photopolymer from the top and the solidified part moves down inside the liquid layer after layer. This orientation allows to make big parts and the biggest ever plastic part possible in AM (Materialise 2020) because the process in this orientation does not defy gravity. But, what happens if the process is turned upside down – the laser beam will come from the down, the photopolymer will be required to be constrained from spilling, the substrate has no room to go down but has only room to go up and consequently the solidified part has to move up layer after layer. In this orientation (Fig. 8.2b), laser beam enters through glass window and solidifies the

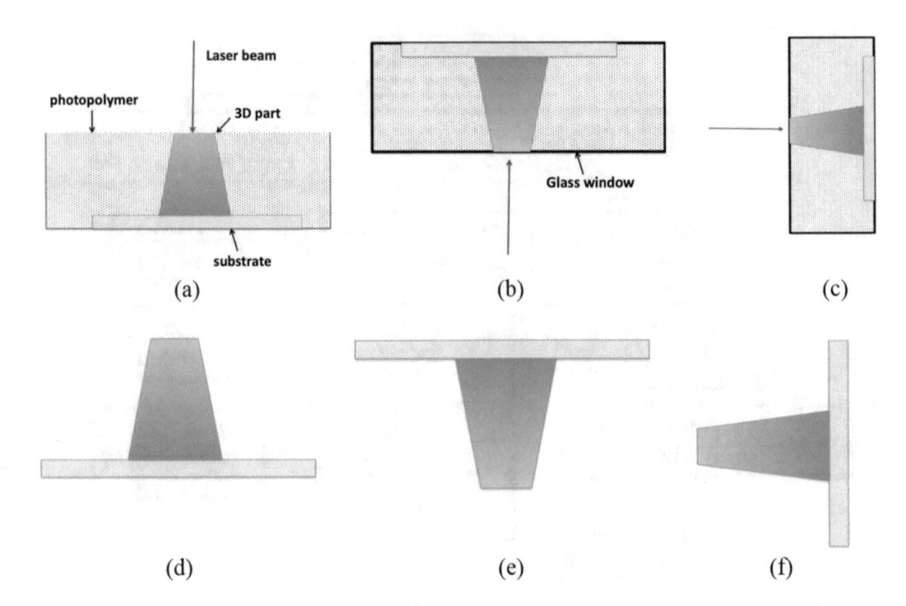

Fig. 8.2 Various orientations of photopolymer bed process (from **a–c**) and the orientation of fabrication of parts therefrom (from **d–f**): (**a**) PPBP in general orientation, (**b**) PPBP in inverse orientation, (**c**) PPBP in sidewise orientation, (**d**) part fabricated in general orientation, (**e**) part fabricated in inverse orientation, (**f**) part fabricated in sidewise orientation

liquid present between a substrate and the glass. In order to prevent the solidified material to be attaching on the glass, a Teflon film is coated on the glass. Successive layers are formed by moving the substrate up and filling up the gap between the solidified layer and the glass by recoating. This orientation may not allow to make a big part but facilitate to make a part with minimum supply of liquid enough to fill the gap between the window and the on-going solidified part (Chi et al. 2013) (given in Chap. 2). Other orientations are also possible such as sidewise orientation (Fig. 8.2c) (Hafkamp et al. 2017). Different orientations allow parts to grow in different orientations as shown in Fig. 8.2d–f (Santoliquido et al. 2019).

8.3 Why Stereolithography Is Not Stereolithography

The name stereolithography has come from lithography, which is a printing process. In lithography, a perforated mask is made through which printing takes place on a substrate coated with a special material. Depending upon the type of image required on a substrate, a mask of such type of image is accordingly created. Thus, lithography is a process to create coating by transferring an image from a mask (Levinson 2005). Stereo means 3D representation. Combining the meaning of stereo and lithography, stereolithography means a process to create a 3D part by transferring an image from a mask. But, this is not what stereolithography does. In stereolithography, photopolymer layers are created and they are solidified as a typical process practised in layer upon layer process (Salonitis 2014). Thus, what stereolithography does is neither an extension of lithography nor having any connection whatsoever with lithography. Thus, the name stereolithography does not represent what the process stereolithography does.

But, since lithography uses a mask and mask based stereolithography uses a mask, this brings a question whether any connection can be found between lithography and stereolithography. Mask in lithography is not same as mask in stereolithography. While mask in lithography is a physical tool, mask in stereolithography is a scanning method. While mask in lithography always represents a constant design, mask in stereolithography may represent variable designs with a change in layer. While mask in lithography works independent of the computer which created design on the mask, mask in stereolithography is controlled by a computer to create part as per design. While presence of mask in lithography confirms that lithography is not a toolless manufacturing process, the presence of mask in stereolithography does not change its status of being a toolless manufacturing process. Thus, the mask is not able to mask the fact that there is no relevant connection between lithography and stereolithography. Henceforth, the name photopolymer bed process (PPBP) is used for stereolithography and its variants (details in Chap. 2).

8.4 Liquid Deposition Process

Liquid deposition process (LDP) is known as ink jet printing (Derby 2015), digital ink jet printing (Lee et al. 2018), direct inkjet printing (Cappi et al. 2008), polymer jetting, photopolymer jetting (Fig. 8.3) etc., it is mainly deposition of low-viscous liquids using a nozzle to make mainly small parts. Material to make a part is provided in the form of liquid through a nozzle, if material is not in the form of a liquid then it is melted so that it could have low enough viscosity to be deposited. If material is of high melting point, such as ceramic, then it could be mixed in a carrier

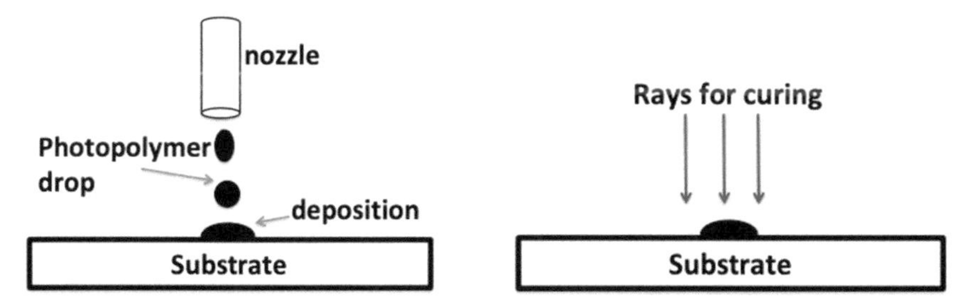

(a) Photopolymer drop is deposited on a substrate by a nozzle

(b) Deposited photopolymer is solidified by curing by rays afterwards

Fig. 8.3 Schematic diagram for photopolymer jetting: (**a**) photopolymer jetted, (**b**) jetted photopolymer solidified

liquid or if material is of high molecular weight polymer then again it is dispersed in a carrier liquid so that ultimately a low-viscous liquid can be obtained which can be ejected through a nozzle. These liquids should not be solidified in the nozzle or during flight so that they could reach to the substrate in the state of a liquid. Besides, the composition of a liquid should not change due to chemical reaction or physical segregation or evaporation before it reaches the substrate. Liquids can reach to the substrate either in the form of a continuous liquid or drops, but continuous liquids have propensity to break before it reaches, therefore, liquid deposition in LDP is usually drop wise deposition. The process can thus be controlled by controlling the ejection of drops and number of drops. Drops are ejected or detached from the bulk of liquid through nozzle by many ways – by application of electric field (Ball et al. 2018), variable magnetic field (Jayabal et al. 2018; Simonelli et al. 2019), vibration, application of sound waves, heating, by application of vapour pressure, by application of force using physical objects such as piston or screw etc. The number of drops ejected is synchronized with the speed of nozzle relative to a substrate, the speed should not be high to cause gap between two deposited drops while the speed should not be low to cause high overlap between two deposited drops causing a decrease in precision.

When a drop reaches to a substrate, the condition of the drop depends upon the impact, if it has high impact, the drop will be flat while its shape will be elliptical. In case of high impact, there will be splash causing a loss of materials from the drop. The elliptical drop will may retract depending upon the high surface tension it has (Derby 2015). If it has low surface tension, the drop will be flatter. In case of reaching of successive drops, the drops will coalesce and the drops will lose their identities to give way a line. Solidification of drops or lines depends upon the phase transformation of drops or polymerization (Fig. 8.3b) and gelling or vaporization of carrier liquid. Solidification can be facilitated by changing the environments on the substrate such as localized heating or irradiation by beam. The timing of solidification needs to be optimized so that the drop should not solidify completely before it coalesces with the next drop in order to become a line. If there are many nozzles then many such adjacent lines can be created in order to make a layer, and thus the fabrication can be expedited. If there are many nozzles using many liquid materials, then there will be many lines made from different materials. Using many nozzles can provide a technique to create a product consisting of many materials; this can be rightly called multi-material product. If many materials intend to serve different functions of a given part or a given product then the product can work as a multi-functional product. But, making a multi-material or multi-material part requires compatibility between different materials so that their ejection from the nozzles will not bring incompatibility in ejection times and drop sizes, besides, there will not be large difference between inter-materials (drops from different nozzles) coalescence and intra-materials (drops coming from the same nozzle) coalescence. If drops coming from different nozzles are of different colours then a multi-colour part can be formed (Meisel et al. 2018).

8.4.1 Water Deposition

What if water is deposited from a nozzle on a substrate and a product is formed, the product made from water will be none other than a water product, that is an ice structure (Barnett et al. 2009). For making such structure, water needs to be deposited in a controlled manner and the substrate needs to be kept at a sub-zero temperature so that the moment water touches the substrate, it gets solidified (Fig. 8.4). This type of solidification is not alien to AM, majority of AM processes go through such solidifications if not from liquid to solid then from solid to liquid to solid. The novelty of this water based solidification is that it is not taking place in a usual environment or at room temperature but is taking place in an environment or it must take place in an environment which requires extra effort to maintain such environment and different materials to work with.

When the ice structure is going to be made on the substrate and the structure is growing, it will start to melt because the substrate temperature though maintained at liquid nitrogen (−140 °C) will not be able to cool it; this requires the whole setup of nozzle and substrate to be kept inside a cool chamber (−20 °C) so that structure will not melt any more (Leu et al. 2009). But, it may affect the water inside the nozzle or water in flight from the nozzle to the substrate; the water will start to solidify. Therefore, the nozzle is kept near the substrate so that transit time is not high enough to be able to convert the water before it reaches the substrate. In order for the water to come smoothly from the nozzle, the water is ejected using more than a critical pressure so that flow of water will not be obstructed by icing on the nozzle, flow of water will break off the icing (Barnett et al. 2009).

A nozzle is not enough when a complex structure having overhanging components is desired to be fabricated, another nozzle is required which will create support structure using another material. Thus, two nozzles are required: one nozzle for main material and other for supporting material. One nozzle can do the job of two nozzles: once it will deposit main material and next time it will deposit supporting material as per need – but then the same nozzle needs to be fitted with different liquid sources periodically which will delay the fabrication, contaminate the liquid, change the heat transfer time and compromise the accuracy. Even if only one nozzle is present instead of two nozzles, two materials are required: one is main material which is water and the second is the supporting material which needs to have lower freezing point than water (such as NaCl solution or brine solution) (Barnett et al.

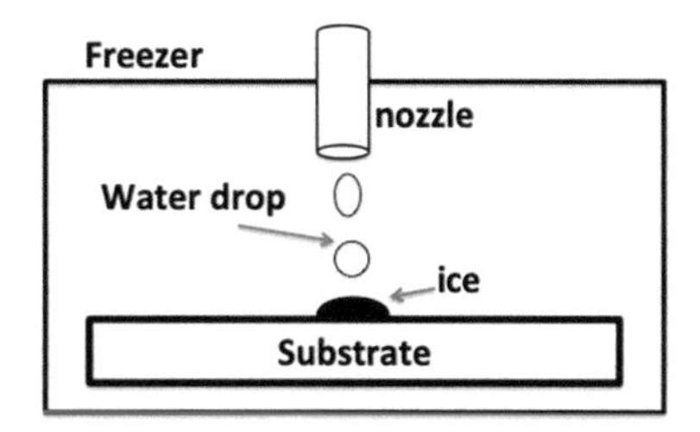

Fig. 8.4 Schematic diagram of rapid freeze prototyping

2009) so that after the fabrication supporting structure made by supporting material can be removed by placing the part in a chamber having temperature more than the freezing point of brine and less than the freezing point of water. Thus, in that chamber, supporting structure will melt away leaving behind the main structure. In an absence of two materials, what will happen if one material (water) will do the job of two materials: water can do the job of supporting material – the supporting structure will be made weaker than the main structure so that the supporting structure can be removed by application of force without damaging the main structure – this method is already used in other AM processes but this method may not be convenient and better because the removal needs to be executed inside a cold chamber.

The ice structure thus fabricated will melt away if removed from a cold chamber and therefore its application needs to be worked out before it is removed. The process can be applied to make an ice pattern for investment casting. In comparison to a polymer pattern, an ice pattern gives an advantage that it does not need to be burnt to be removed from a ceramic shell, it is just removed by bringing the pattern (covered with ceramic shell) outside the cold chamber. At room temperature, ice pattern completely melts away without leaving any residue and thus a ceramic shell is formed for further use (casting). In order to create a ceramic shell, the ice pattern thus fabricated needs to be dipped in a ceramic slurry. The slurry should not be frozen so it requires to be free from water and thus a special slurry is required and thus a material which works in a usual environment may not work in sub-zero environment. The slurry coated on the pattern dries with the help of a catalyst and the shell thus formed from the slurry is further strengthened by usual heat treatment (Zhang and Leu 2000).

Ice parts formed from this process have not many applications, if possible applications in the area of making ice sculpture are not counted, but the method learnt is applied in an area of tissue engineering where a scaffold not printed in a sub-zero environment will be affected by porogens. Thus, scaffolds are printed using this method by replacing water with a biological solution – this process is named as cryogenic prototyping (Pham et al. 2008) while the process using water is known as rapid freeze prototyping (RFP) (Bryant et al. 2003).

8.5 Slurry Based Process

Slurry is a mixture of solid and liquid in which solid is suspended in the liquid. In order for the solid to remain suspended, dispersant is required which gets adsorbed on the surface of a solid particle and does not let two solid particles to come in contact with each other – this causes solid particles not to be agglomerated and remain dispersed in a liquid medium. Increasing dispersant is counterproductive as it decreases dispersion and increases agglomeration by bonding solid particles with dispersant polymer chains. The liquid medium is called solvent which may be water or other polymeric solutions (ethanol, acetic acid etc.), water is not suitable for some materials which get oxidized or solidified in reaction with water (Wu et al.

2019). Slurry may also require plasticizer, the role of the plasticizer is to increase fluidity of the slurry or to decrease viscosity of the slurry. Slurry may also require binder, the role of the binder is to hold slurry particles together when the slurry is dried. Slurry is used in AM in two ways: (1) by spreading slurry on a platform similar to powder bed process, henceforth named as slurry bed process, and (2) by depositing slurry on a platform similar to fused deposition modelling or material extrusion process, henceforth named as slurry deposition process.

8.5.1 Slurry Bed Process

Slurry bed process has following advantages over powder bed process:

1. It is not convenient to make a layer using sub-micron sized powders as the powders repel each other due to electrostatic force. Using slurry made from these powders helps make a thin layer, not achievable in powder bed process.
2. Powder bed process does not work well with powders having broad size distribution, non-spherical powders, powders having high difference in density while spreading of slurry is not affected by these powder properties.
3. Powder layer possesses lower density; using slurry instead of powder will increase the density. Initial higher density of the layer helps achieve higher final density of a part.

In this process, with the help of a doctor blade, slurry is laid on a bed which is then dried so that deposition of the next layer of slurry should not deform it (Fig. 8.5). In case of an aqueous based slurry, drying can happen by raising the substrate temperature to more than the boiling point of water. Without increasing the temperature, layer can also dry by capillary action when the solvent will be absorbed by previous layers (Muhler et al. 2015). The dried layer is selectively treated by a laser beam (Muhler et al. 2015) or a binder jet (Zocca et al. 2019) to create a pattern on it as per

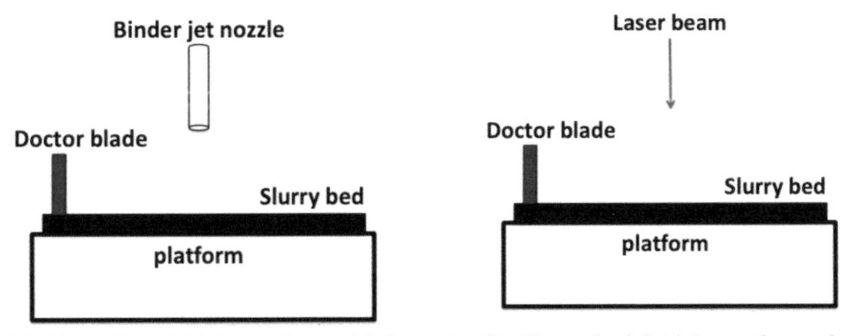

(a) Slurry bed laid by a doctor blade and shaped by jetted binder

(b) Slurry bed laid by a doctor blade and shaped by a laser beam

Fig. 8.5 Schematic diagram of slurry bed process: (**a**) shaping due to binder jetting, (**b**) shaping due to scanning by a laser beam

a CAD file. The processed part, also known as a green part, may go through post-processing to improve its properties (Wang et al. 2004).

Viscosity of a layer is an important parameter in a slurry bed process. For smooth spreading of the layer, viscosity needs to be small. But, if viscosity is small, there will be higher liquid content in the slurry which needs to be removed during drying – this may give rise to cracks in the slurry. Low viscosity may also imply that there is no high content of solid particles which will give rise to low density of the layer and subsequently of the part. Low density means low strength which may not be desired unless a porous part is desired for some applications.

8.5.2 Binding Methods in Slurry Bed Process

A dried slurry layer is itself a layer having bonded particles but this bonding will dissolve during post-processing. This layer needs to be additionally treated, in the same way as treated in powder bed process, to make a pattern on it by creating additional bonds among slurry particles at selected areas – these areas being constituent of a desired part will survive during post-processing dissolution. Additional bond is created by a laser beam, same as in SLS /SLM, when the beam partially melts slurry and may vaporize polymers – this process is known as laser slurry deposition (Muhler et al. 2015) or ceramic laser fusion (CLF) (Tang 2002). The bond is also created by jetted binder same as in BJ3DP, when the binder fills in pores and locks separate particles, and permeates through the slurry layer to bond with an underlying layer – this process is known as laser slurry deposition-print (Lima et al. 2018).

Bonding can also be created by activating binders present in a non-dried slurry layer; colloidal silica (sol) present in the layer remains dormant unless the layer is scanned by a laser beam, the beam converts sol into gel, which bonds adjacent particles. This process is known as selective laser gelling (Liu and Liao 2010; Liu et al. 2013). Thus, there are three shaping or binding methods in slurry bed process – binder jetting (Fig. 8.5a), laser melting (Fig. 8.5b) and laser gelling as shown in Fig. 8.6.

Slurry mixed with photopolymer can also be spread in the form of a bed to make a layer. Since the solid content (about 40 vol.%) in photopolymer is low, the process is nearer to photopolymer bed process rather than slurry bed process. This type of slurry can be conveniently processed akin to photopolymers. Low solid content means low deflection of laser beam which allows adequate curing thickness, more

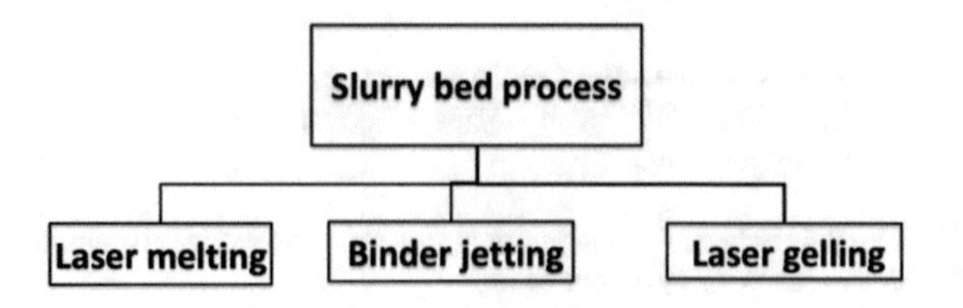

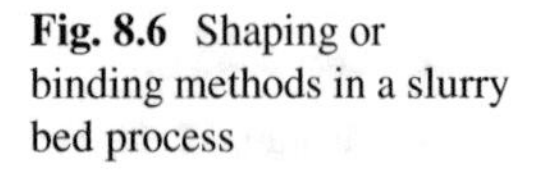

Fig. 8.6 Shaping or binding methods in a slurry bed process

than the layer thickness, to be obtained to ensure sufficient bonding between two slurry layers. The method of bonding is cured by a laser beam, similar to setereo-lithography. This process is called lithography based ceramic manufacturing (LCM) (Harrer et al. 2017; Schwarzer et al. 2017).

8.5.3 *Slurry Deposition Process*

In slurry deposition process, slurry is deposited from a nozzle to make 3D structures; this helps to make ceramic and metal components (Fig. 8.7). If slurry is not used, there are no other ways to make ceramic or metal components in a nozzle based process but by melting them. Melting high melting point materials is not so easy; it requires devices for melting, it requires dedicated systems for heat management, it requires to control microstructure development and crack mitigation – these all will increase cost as well (Sames et al. 2016). Thus, there are expensive systems which are powder based or wire based requiring laser or electron beam (Frazier 2014). Melting low melting point materials such as a tin based alloy (Vega et al. 2014) or an aluminium based alloy (Zuo et al. 2016) is not so difficult; there are processes and systems available for depositing these materials (Fang et al. 2017) – but parts made from these materials are weak, they cannot be substituted for a high strength part; the process is incapable as it cannot be used for processing high melting point materials. A slurry based deposition process comes up as a solution to these problems – it is neither as expensive and difficult as a nozzle based process related to high melting point materials nor as incapable as a nozzle based process related to low melting point materials.

Slurry deposition process can be a substitute for nozzle based processes related to high melting point materials but slurry is not a substitute for these materials. Slurry contains these materials and it requires a carrier to carry these materials; the carrier may be water or some polymers. The amount of these solid materials is maximum around 30 vol% in ink jetting process (Cappi et al. 2008; Özkol et al. 2009; Hagen et al. 2019) and maximum around 60 vol% in 3D gel printing (Ren et al. 2016). Lower the amount means lower disturbance to the original process, for example if solid content in ink jetting process is just 2 vol%, the process parameters would not be much different than that required for pure ink, thus less obstacle to the

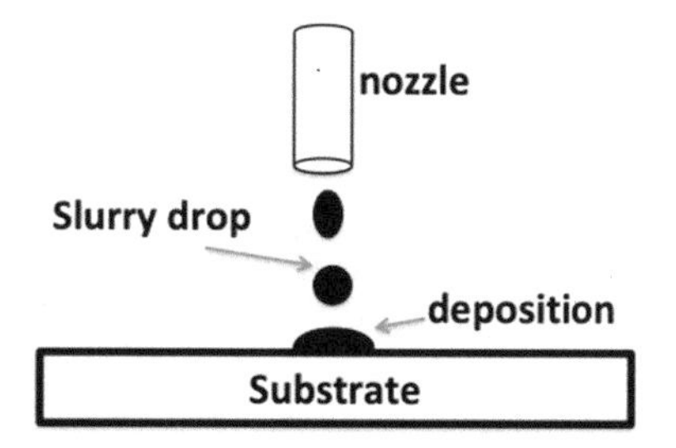

Fig. 8.7 Schematic diagram of slurry deposition process

ink flow. But, such low amount would not help getting fully dense materials and may not be interesting for many applications.

Slurry extruded on a platform should not be deformed on its own weight and sustain the form when another layer is deposited on it; higher solid loading is required to provide strength to the extruded material. However, very high solid loading requires high force to extrude and the continuity of the extruded material may break, which will give rise to defects in part fabrication (Tang et al. 2019). Extruded materials are bonded with other extruded materials or previously deposited layer before they get solidified; bonding is due to the binder or gel present in a slurry. Gel is preferred in some materials (tricalcium silicate) because it prolongs solidification of extruded filaments, which gives filaments sufficient time to bond; gel also increases flow properties of the slurry which helps the material to be extruded uniformly (Wu et al. 2019).

8.6 Four Dimensional Printing

Parts made from AM go through post-processing. It is mainly done due to two reasons:

1. To improve properties so that parts could become fit to be usable. This post-processing ranges from intensive cleaning to treatments lasting for several hours or days. This post-processing aims to compensate what could not be achieved during processing. If AM process is developed well, then this post-processing will no longer be required. For example, if a polymer part made from stereolithography is not cured then post-curing as post-processing is applied to fully cure the polymer part. If stereolithography in that case is working well, then there is no need for post-processing. Since the process (stereolithography) did not give well-cured parts, post-processing is done to compensate the demerit of the process.
2. To achieve properties that are not achieved from AM not because AM is not well developed for a particular type or particular material but because product requirement is different, for example, post-processing as coating with hydroxyapatite to improve biocompatibility of a polymer part made from fused deposition modelling (FDM). In this case, post-processing is required not because FDM could not furnish good coating but because it was not expected from FDM to have any such coating like properties, it was not the job of FDM to furnish good coating or any coating. Post-processing is not done to compensate a sub-standard part (made from a process) but because the best part (made from the best process) was not the requirement of the product.

AM plus post-processing is as old as AM but the aim of post-processing has always been to improve properties. Four-dimensional printing (4DP) (Momeni et al. 2017) is a relatively new technique which is also AM plus post-processing but the aim of the post-processing is to change the shape rather than to change properties. It does not mean that there will be no property changes in 4DP, properties will change but

Fig. 8.8 Schematic diagram of 4DP

changing the property is not the primary aim of 4DP; if there is no shape change in 4DP, then it will not be called 4DP even if there is a change in properties.

4DP starts with the fabrication of an AM part, the part is then post-processed to change its shape (Fig. 8.8). AM processes used are photopolymer based processes, filament deposition, ink deposition and powder bed fusion (Shafranek et al. 2018). For post-processing, the part is kept in a certain environment where it changes shape in a controlled way due to the presence of pressure, heat, light or chemicals etc. Since all materials do not change their shapes in an environment, there requires a certain type of materials known as smart materials which will respond to a particular energy when kept in an environment. Since two types of materials may change differently in an environment – if a part is made up of these materials then they will give rise to different shape changes – if they will be treated in different environments for different lengths of time, the number of final shapes obtained will increase. Thus, 4DP is capable to provide a number of shapes obtained from a combination of materials, environments, designs and time periods.

4DP, as the name suggests, has four dimensions – three dimensions for space and an additional dimension for time, while an object of three dimension is acquired by 3DP (or AM), neither the object nor its shape is final until the application of the fourth dimension is exercised. 4DP is a generic process which starts from various AM processes such as stereolithography, ink jet printing, fused deposition modelling etc., and is thus not an AM process nor a variant of AM process but is an application of AM processes. 4DP has been defined as 'additive manufacturing of objects able to self-transform, in form or function when are exposed to a predetermined stimulus, including osmotic pressure, heat, current, ultraviolet, or other energy sources'. The definition of 4DP does not disagree with the fact that 4DP is not an AM process. As per the definition of 4DP, 4DP is the recipient of objects fabricated by AM. When objects are fabricated by AM, then there are many recipients: when these are fabricated for aerospace applications then aerospace industry is a recipient, when these are fabricated for automotive applications then automotive industry is a recipient. But, these recipients, like 4DP, are but an AM process.

References

Ball AK, Das R, Das D et al (2018) Design, development and experimental investigation of E-jet based additive manufacturing process. Mater Today Proc 5:7355–7362

Barnett E, Angeles J, Pasini D, Sijpkes P (2009) Robot-assisted rapid prototyping for ice sculptures. In: IEEE International Conference on Robotics and Automation, Kobe, Japan

Bertsch A, Lorenz H, Renaud P (1999) 3D microfabrication by combining microstereolithography and thick resist UV lithography. Sensors Actuators A Phys 73(1–2):14–23
Bryant FD, Sui G, Leu MC (2003) A study on effects of process parameters in rapid freeze prototyping. Rapid Prototyp J 9(1):19–23
Cappi B, Özkol E, Ebert J, Telle R (2008) Direct inkjet printing of Si3N4: characterization of ink, green bodies and microstructure. J Eur Ceram Soc 28(13):2625–2628
Chen Y, Zhou C, Lao J (2011) A layerless additive manufacturing process based on CNC accumulation. Rapid Prototyp J 17(3):218–227
Chi Z, Yong C, Zhigang Y, Behrokh K (2013) Digital material fabrication using mask-image-projection-based stereolithography. Rapid Prototyp J 19(3):153–165
Derby B (2015) Additive manufacturing of ceramic components by ink jet printing. Engineering 1(1):113–123
Fang X, Wei Z, Du J et al (2017) Forming metal components through a fused-coating based additive manufacturing. Rapid Prototyp J 23(5):893–903
Frazier WE (2014) Metal additive manufacturing: a review. J Mater Eng Perform 23:1917–1928
Hafkamp T, Baars GV, Jager BD, Etman P (2017) A trade-off analysis of recoating methods for vat photopolymerization of ceramics. SFF Symp Proc 28:687–711
Hagen D, Kovar D, Beaman J J, Gammage M (2019) Laser flash sintering of additive manufacturing of ceramics. ARL-TR-8657, Defence Tech Info Centre, US
Harrer W, Schwentenwein M, Lube T, Danzer R (2017) Fractography of zirconia-specimens made using additive manufacturing (LCM) technology. J Eur Ceram Soc 37:4331–4338
Jacobs FP (1992) Rapid prototyping and manufacturing: fundamentals of stereolithography. Society of Manufacturing Engineers, Dearborn
Janusziewicz R, Tumbleston JR, Quintanilla AL et al (2016) Layerless fabrication with continuous liquid interface production. PNAS 11(42):11703–11708
Jayabal DKK, Zope K, Cormier D (2018) Fabrication of support-less engineered lattice structures via jetting of molten aluminum droplets. In: SFF Symposium Proceedings, pp 757–764
Lee JH, Kweon JW, Cho WS et al (2018) Formulation and characterization of black ceramic ink for a digital ink-jet printing. Ceram Int 44:14151–14157
Leu M, Isanaka SP, Richards VL (2009) Increase of heat transfer to reduce build time in rapid freeze prototyping. In: SFF Symposium Proceedings, pp 219–230. www.materialise.com
Levinson HJ (2005) Principles of lithography. SPIE, Bellingham
Lima P, Zocca A, Acchar W, Günster J (2018) 3D printing of porcelain by layerwise slurry deposition. J Eur Ceram Soc 38(9):3395–3400
Liu FH, Liao YS (2010) Fabrication of inner complex ceramic parts by selective laser gelling. J Eur Ceram Soc 30(16):3283–3289
Liu FH, Lee RT, Lin WH, Liao YS (2013) Selective laser sintering of bio-metal scaffold. Proc CIRP 5:83–87
Meisel N, Dillard D, Williams C (2018) Impact of material concentration and distribution on composite parts manufactured via multi-material jetting. Rapid Prototyp J 24(5):872–879
Momeni F, Hassani SMM, Liu X, Ni J (2017) A review of 4D printing. Mater Des 122:42–79
Muhler T, Gomes C, Ascheri M et al (2015) Slurry-based powder beds for selective laser sintering of silicate ceramics. J Ceram Sci Technol 06(02):113–118
Özkol E, Ebert J, Uibel K et al (2009) Development of high solid content aqueous 3Y-TZP suspensions for direct inkjet printing using a thermal inkjet printer. J Eur Ceram Soc 29(3):403–409
Pham CB, Leong KF, Lim TC, Chian KS (2008) Rapid freeze prototyping technique in bio-plotters for tissue scaffold fabrication. Rapid Prototyp J 14(4):246–253
Ren X, Shao H, Lin T, Zheng H (2016) 3D gel-printing- an additive manufacturing method for producing complex shaped parts. Mater Des 101:80–87
Rudraraju A, Das S (2009) Digital date processing strategies for large area maskless photopolymerization. In: SFF Symposium Proceedings, pp 299–307
Salonitis K (2014) Stereolithography. Compr Mater Process, Elsevier 10:19–67

Sames WJ, List FA, Pannala S et al (2016) The metallurgy and processing science of additive manufacturing. Int Mater Rev:1–46
Santoliquido O, Colombo P, Ortona A (2019) Additive manufacturing of ceramic components by digital light processing: a comparison between the "bottom-up" and the "top-down" approaches. J Eur Ceram Soc 39(6):2140–2148
Schwarzer E, Götz M, Markova D et al (2017) Lithography-based ceramic manufacturing (LCM) – viscosity and cleaning as two quality influencing steps in the process chain of printing green parts. J Eur Ceram Soc 37(16):5329–5338
Shafranek RT, Millik SC, Smith PT et al (2018) Stimuli-responsive materials in additive manufacturing. Progr Polym Sci 93:36–68
Simonelli M, Aboulkhair N, Rasa M et al (2019) Towards digital metal additive manufacturing via high-temperature drop-on-demand jetting. Addit Manuf 30:100930
Tang HH (2002) Direct laser fusing to form ceramic parts. Rapid Prototyp J 8(5):284–289
Tang S, Yang L, Li G et al (2019) 3D printing of highly-loaded slurries via layered extrusion forming: parameters optimization and control. Addit Manuf 28:546–553
Vega EJ, Cabeza MG, Muñoz-Sánchez BN et al (2014) A novel technique to produce metallic microdrops for additive manufacturing. Int J Adv Manuf Technol 70:1395–1402
Wang HR, Cima MJ, Kernan BD, Sachs EM (2004) Alumina-doped silica gradient-index (GRIN) lenses by slurry-based three-dimensional printing (S-3DP™). J Non-Cryst Solids 349:360–367
Wu W, Liu W, Jiang J et al (2019) Preparation and performance evaluation of silica gel/tricalcium silicate composite slurry for 3D printing. J Non-Cryst Solids 503–504:334–339
Zhang W, Leu MC (2000) Investment casting with ice patterns made by rapid freeze prototyping. In: SFF Symposium Proceedings, pp 66–72
Zhou X, Hou Y, Lin J (2015) A review on the processing accuracy of two photon polymerization. AIP Adv 5:030701
Zocca A, Lima P, Diener S et al (2019) Additive manufacturing of SiSiC by layerwise slurry deposition and binder jetting (LSD -print). J Eur Ceram Soc 39(13):3527–3533
Zuo H, Li H, Qi L, Zhong S (2016) Influence of interfacial bonding between metal droplets on tensile properties of 7075 aluminum billets by additive manufacturing technique. J Mater Sci Technol 32(5):485–488

Chapter 9
Air and Ion Deposition Processes

Abstract For fabricating a part of size some 100 microns or few millimetre, a resolution of nanometre is required. There are some processes such as aerosol jetting and electrochemical additive manufacturing (AM) which add ions by ions or atoms by atoms to make parts with a possibility to have nanometre resolution. These processes have potential to be applied in the area of fabricating electronic lines and parts having micropores. This chapter describes these air based and ion based processes and explains why electrolytic solution based AM processes have potential to overcome manufacturing problems posed by layer upon layer processes.

Keywords Aerosol jetting · Electrochemical · Electrolytic solution · Build direction · Ions · Colloids

9.1 Aerosol Jetting

The position of aerosol jetting along with ions based AM processes in AM classification is given in Fig. 9.1.

Aerosol jetting (AJ) or aerosol jet printing (Goh et al. 2018) is an air based deposition process, which is used to fabricate small features, mainly for electronic applications (Wilkinson et al. 2019).

Aerosol implies that liquid or solid particles are suspended in air; mainly liquid particles are used in AJ – these particles become constituent of a part made by AJ while air acts as carrier to transport these particles and facilitate the process. If particles are suspended in air then they are not usually big, they are of the size of nano or micrometre; if such small particles need to be transported by air, then there will be a small amount of particles per second ready to be deposited and make a part. Thus the process is suitable to make only small parts from ~10 μm up to 1 mm with ~100 nm resolution. Most of the parts are used along with substrates on which the parts are made; the process is thus used more for modifying a substrate and adding value to the substrate than making an independent part to be removed from the substrate. Most of applications of the process are creating structures such as antenna,

S. Kumar, *Additive Manufacturing Processes*,
https://doi.org/10.1007/978-3-030-45089-2_9

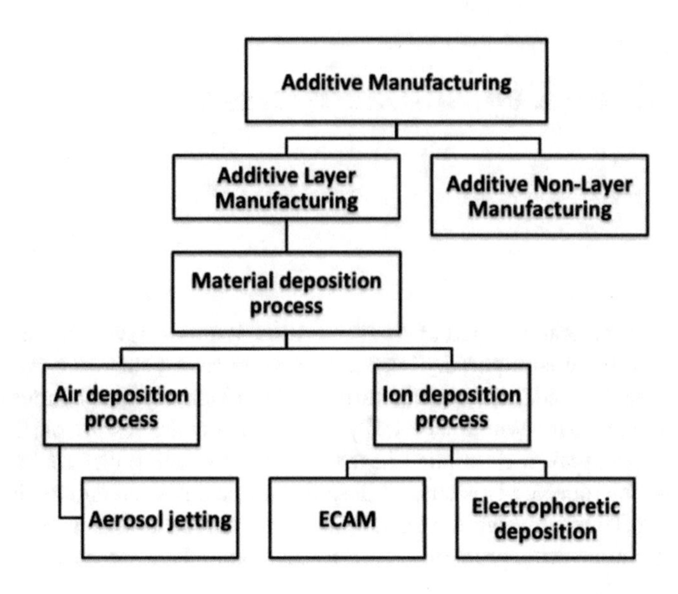

Fig. 9.1 Position of air and ion based processes in AM classification

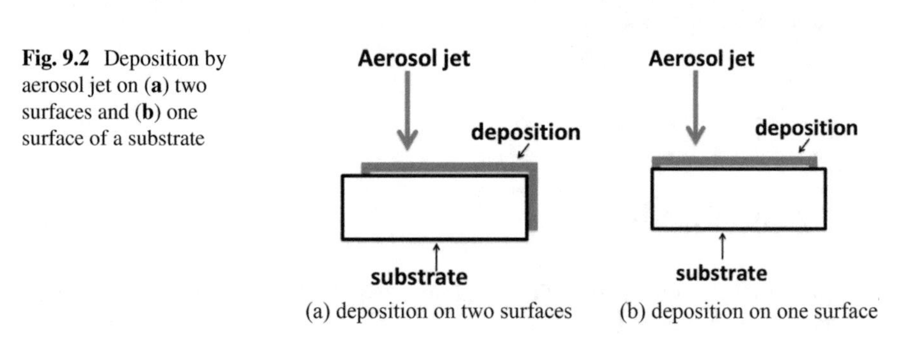

Fig. 9.2 Deposition by aerosol jet on (**a**) two surfaces and (**b**) one surface of a substrate

interconnects, electronic circuits, conductive lines either on planar or non-planar surfaces – these fabrications, even if made on 3D substrates, may not qualify this process as AM if these are not multi-layer. An example of a 3D substrate is a cube where deposition by AJ is going to be done at least on two of its six surfaces (Fig. 9.2a). While an example of a 2D substrate is a cube where deposition by AJ is going to be made on only one of its six surfaces (Fig. 9.2b). Thus, as per definition of AM (ASTM 2012), multi-layer coating deposited by AJ as shown in Fig. 9.2a is a coating and not an AM part.

For making a structure, AJ requires creation of aerosol and its subsequent deposition (Fig. 9.3). Creation of aerosol requires creation of small liquid droplets and mixing them with air. Creation of small liquid droplets requires agitation of liquid so that separation of droplets from the bulk liquid takes place. Agitation of liquid is possible by vibrating a mass of liquid placed in a container so that due to vibration some liquid particles on the surface of the liquid overcome surface tension force and

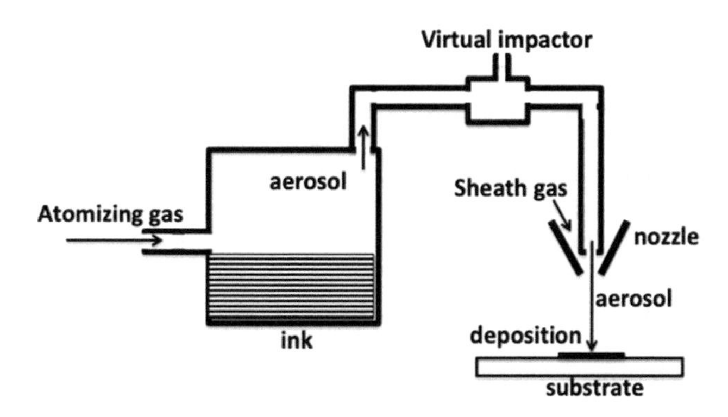

Fig. 9.3 Schematic diagram of aerosol jetting

get separated from the bulk to form droplets. This is also the concept of ultrasonic atomization where ultrasonic vibration is transferred to the liquid through some other liquid medium.

Agitation of liquid is also possible by impacting it with a gas moving with high velocity, generally coming from a compressed source. Such gas moving parallel to a liquid surface and brushing the surface will detach some liquid from the surface turning the liquid into small droplets. If the viscosity of the liquid is high, higher velocity is required. Alternatively, a high-viscous liquid needs to be heated so that the liquid will be internally agitated and the same gas velocity will give increased amount of droplets. Heating as a method to facilitate droplet formation is generally used when liquid kept still in a container needs to be atomized. If a high-velocity gas moves perpendicular to a liquid stream coming from a nozzle, then collision of the gas with the liquid stream will create droplets – this method is used for atomizing such liquid streams.

Liquid droplets thus created will fall and get lost if these are not carried away. The gas, which creates them, called as an atomizing gas, can also carry them, can then be called a carrier gas. After atomizing, droplets of various sizes are created; higher size (or of higher inertia) droplets will require higher velocity carrier gas in order to be carried away; if the atomizing gas is a carrier gas then the option for increasing or decreasing the velocity is limited. There requires some mechanism after the creation and transportation of aerosols, and before their deposition to exclude droplets of extreme (high or low) size – such droplets are not suitable if a high resolution is required.

When aerosol is moving in a pipe or a tube, creation of an exhaust or outlet in the pipe may let some gas (containing aerosol) to move out to atmosphere through the exhaust; the amount of the gas moving out depends upon the exhaust – if the size of the exhaust is big or the exhaust flow rate is high, more gas will move out resulting in a loss of smaller size droplets to the atmosphere. The effect of exhaust is realized not only in the loss of some droplets but also in the decrease of the gas velocity. If the gas velocity decreases, it will not be able to carry big size droplets and they will eventually be dropped from the gas. Thus, using an exhaust eliminates both smallest and biggest droplets from a carrier gas. This mechanism or this controlling device is called a virtual impactor, which eliminates from the carrier gas extreme size droplets because of their lowest and highest inertia.

Carrier gas deprived of droplets of extreme sizes and containing medium size of droplets is directed towards a nozzle to be deposited through it. If the gas will be sent through the nozzle then the size of a gas flow diameter will increase with an increase in the distance of the nozzle from the substrate – the gas will not provide patterns with high definition and high density. In order to collimate a carrier gas, another gas, called sheath gas, is used which flows in the direction of the carrier gas flow, covering it with an aim to converge at some point – if the carrier gas is just like a solid cylinder then the sheath gas is a hollow converging cone surrounding the cylinder. The sheath gas collimates the carrier gas and helps constrict the deposition. If there is not one carrier gas flow but several carrier gas flows (containing different materials) moving with the same velocity and originating from several different atomizers, then it is possible to merge these flows into one line and make multi-material deposition. Changing the amount of materials in one gas flow by changing the setting of an atomizer will provide more variations in deposited materials.

9.2 Ionic Solution Based Additive Manufacturing

When an external potential difference is applied across a solution using electrodes, charged particles or ions move through the solution – positive ions or cations will move towards a negative electrode called cathode while negative ions or anions will move towards a positive electrode called anode. Movement of ions towards electrodes may lead to accumulation of materials or deposition of materials on electrodes – this is akin to deposition of materials on a substrate in AM – this is a method that is used to add materials in an ionic solution based additive manufacturing. This method is not new, it is already practised in electroplating or electrophoretic deposition for making thin films or coatings, which are 2D structures. The method is also applied in a process named electroforming to make 3D structures (Castellano et al. 2017). In electroforming, materials are deposited on electrodes of various shapes to make structures of various shapes, these electrodes are called a tool or mandrel. On the removal of the electrode from the deposited material, the remaining hollow structure made by the deposited material is a desired 3D structure – this structure is dependent upon the shape and size of a mandrel or tool (Matsuzaki et al. 2019). This dependency of the 3D structure on a tool limits the (electroforming) process to fabricate only those structures which will be conforming to tools; and this dependency does not allow to make a freeform structure. The present AM process is free from such limitations, though it utilizes the similar type of ionic movement to fabricate 3D structures.

Ionic solution implies a liquid either containing ions (e.g. NaCl solution) or having potential to be ionized when an external voltage difference is applied. When potential is applied across an ionic solution such as copper sulphate solution, copper ions move giving rise to copper plating – this happens in electroplating and the solution is also called electrolytic solution; AM process based on these types of solution is an electrolytic solution based AM, also known as electrochemical AM (Kamraj et al. 2016) (Fig. 9.4). A colloidal solution may also be an ionic solution because it contains charged particles, which are suspended in liquid; these charges remain present in the solution even in an absence of external applied electric field. These charged particles move in the presence of external applied electric field and give rise to deposition – this happens in electrophoretic deposition and the ionic solution is colloidal solution; AM process based on these types of solution is a colloidal solution based AM, also known as electrophoretic deposition based AM (Mora et al. 2018).

9.2.1 Electrolytic Solution Based Additive Manufacturing

When metal ions move from an anode towards a cathode and get deposited on the cathode then they form a layer on it. If cathode is a big plate while anode is a small rod facing small area of the cathode, then the formation of the layer will be limited to that area (Fig. 9.4). Exact size of the layer will depend upon various variables such as gap between two electrodes, throwing power of the electrode, applied voltage, form of the voltage, insulation on the side surface of the anode, concentration of electrolytes etc., but changing the size of the anode provides a way to localize the formation of layers or deposition of materials at some selected area (Habib et al.

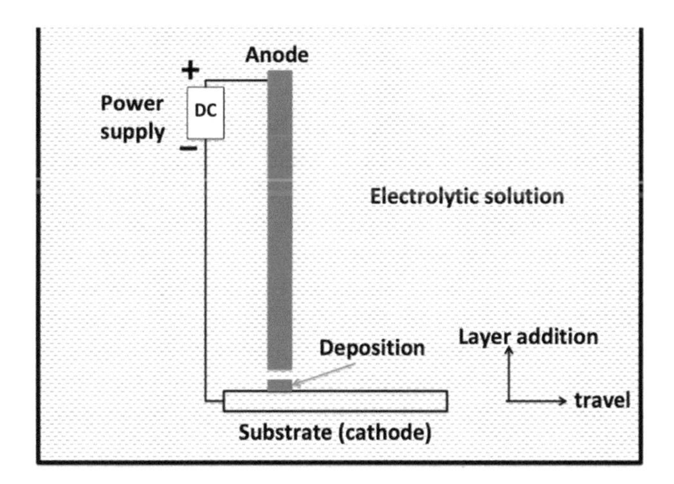

Fig. 9.4 Schematic diagram of electrolytic solution based AM

2009, Lin et al. 2010). If the anode moves parallel to the cathode, then the deposited layer moves and creates a layer in the form of a line having width corresponding to the width of the anode. If the anode moves away from the cathode then the deposited material is no longer confined to a layer but starts making a pillar on the cathode. After the formation of the pillar, if the anode again moves parallel to the cathode, then another pillar will form on the earlier pillar at right angle to it; with a change in direction of the movement of the cathode, many such pillars can be formed – this will result in a structure similar to a pillar having many arms – these are overhang structures which, in other AM processes, require either support structures or change in orientation of the geometry (Paul and Anand 2015) but in this process these are formed without any such inconvenience (Brant and Sundaram 2016). In this process, addition of materials happens ion by ion or atom by atom and when these ions or atoms are getting added on the side surface of a pillar to make an arm then these atoms have no such possibility to succumb to the gravitational force and collapse as happens in a drop by drop AM process. In an AM process where materials are added drop by drop, such as in ink jetting process, liquid drops will need support before they solidify. It does not imply that in this process an overhang structure will never collapse; the structure may collapse if the process of adding ions is not well optimized and the addition does not furnish enough strength to the structure, but the structure will not collapse because ions could not be placed on the side surface to make the structure. While in an ink jetting process or a fused deposition modelling process, formation of the structure, if initiated, will start to collapse because liquid drops or extruded drops could not be placed on the side surface to make the structure.

An anode moving over a cathodic plate will create a line on it made from an electroplated material. Creating a line with such type of movement is not new in AM – a line of material is created on a powder by the movement of a laser beam or an electron beam in powder bed process; a line of material is also created by movement of printer head or nozzle in ink jetting process or fused deposition modelling process, respectively. Addition of many such lines gives rise to formation of a layer and a start for layer upon layer fabrication, and this process is ready to follow such fabrication method. But, this process is also ready not to follow such fabrication method – this process gives opportunity to add materials wherever it is desired to be added or wherever it is desired not to be added, leading a convenient method to make a particular complex part (Manukyan et al. 2019). The process gives opportunity to add materials anywhere on a pillar and make arms at various angles. If a cylindrical pillar having cylindrical arm at right angle to it needs to be fabricated, the cross-sectional area of the pillar is at least ten times the size of the cross-sectional area of a cylindrical tool or anode while cylindrical arm is having five times the cross-sectional area of the tool. If layer upon layer fabrication method is followed, then the pillar can be fabricated well but the arm will show gaps between two layers (Fig. 9.5a); or during the fabrication of the arm, disadvantage of layerwise manufacturing will be visible.

If after the fabrication of the pillar, the tool does not follow such layerwise fabrication method or the tool does not follow such layerwise fabrication method in that build

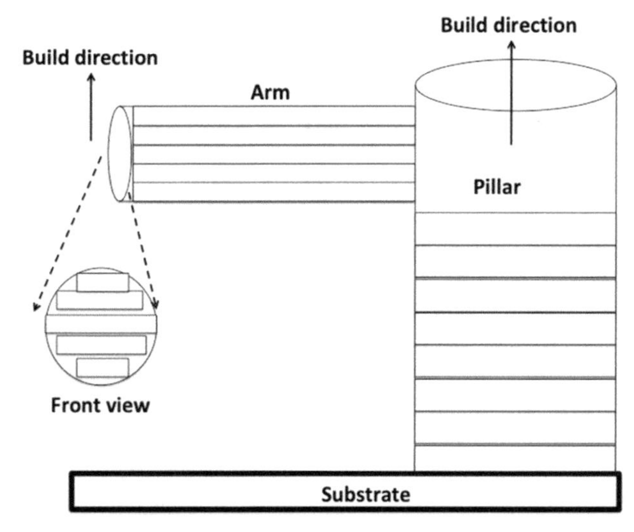

(a) Staircase effect due to same build direction of arm and cylinder

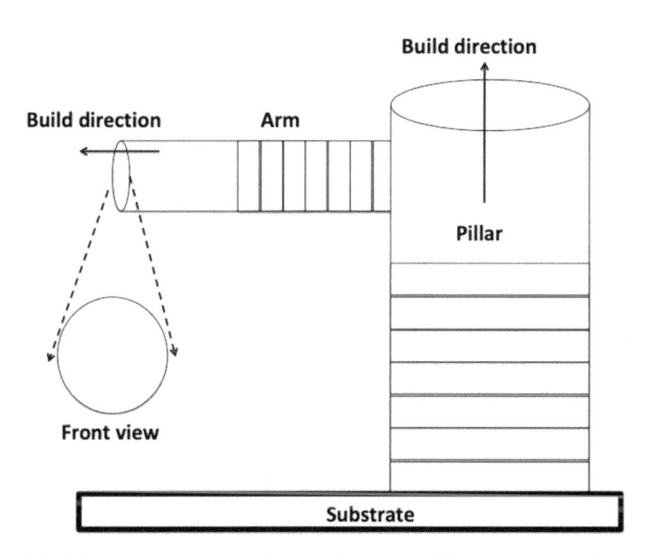

(b) No staircase effect due to different build directions

Fig. 9.5 Effect of build directions in layer upon layer process (given in Chap. 10 as well)

direction, then there will not be such disadvantage of layerwise manufacturing. After the layerwise fabrication of the pillar – if the same story is repeated – the pillar acts as if it is a new cathodic plate, fabrication of arm becomes the fabrication of pillar and the tool still follows the layerwise method – the resulting structure (arm) will not show such non-uniformity and gap noticed earlier because layerwise fabrication method has changed the build direction by 90° (Fig. 9.5b). Thus, the process provides an alternative method to make a part, or the process gives an opportunity to find a convenient method to make a complex part because the process has an advantage to add materials wherever it is desired to be added.

The present example of pillar and arm showed that how the disadvantage of layerwise method can be overcome just by changing the build direction and without any need for relinquishing the method itself. But, just changing the build direction may not be just changing the build direction – there is no software available which will slice half of a CAD model in horizontal direction and the other half in vertical direction. All AM systems have just one build direction (Coupek et al. 2018). Though, it can be doable – (1) after fabrication of a pillar, if the position of the cathodic plate is changed by 90° and the build is continued, (2) after fabrication of a pillar, a new CAD model is used for fabricating the arm; the model is sliced vertically instead of horizontally and a new tool path is created; this new tool path is giving a build direction which is at 90° to the old build direction that was used to create the pillar; the CAD model is positioned at the position of the arm of the old CAD model and the build is continued. Both options show it is not impossible to manufacture one pillar and arm. But what if the part is not so simple, the part does not consist of just one pillar and arm but has hundreds of pillars and arms at various different angles, then the above options will still work but it would not be convenient. In that case, massive process planning will be required – and the advantage gained in the process will be lost in the planning.

There is a clear advantage when build direction is changed but there is a clear problem for changing the build direction. The problem increases when layerwise fabrication needs to be maintained while changing the direction – one of the reasons for increase is that a layer is big and inflexible. If the size of a part is big or the size of its cross-section is big, the size or perimeter of the layer will be big. Similarly, if the size of a part is small, layer size will be small. Perimeter of a layer is inflexible and constant; it comes along with the 3D model of a part to be fabricated. If there is any flexibility, then it is in the thickness of a layer – layer thickness can be varied – it is a candidate open to variation during fabrication. In case of a change in orientation of a part having unequal dimensions, the perimeter can be changed but it is still decided by the model of the part, and is not a candidate open to variation during fabrication, and is therefore constant and inflexible during fabrication.

In layerwise fabrication, it is the layer which is a basic building block – it is not because layer is not made from smaller blocks but because unless a complete layer is fabricated the fabrication refuses to progress to the next layer; there is no such possibility as to delay the fabrication of a fraction of a layer till next two or three layers will be formed. One of the biggest problems of layerwise fabrication is that layer size or perimeter is always very big – it does not imply that the method is only meant to make big parts, it also does not imply that fabrication of small parts is not amenable – it only implies that the layer size is always very big in comparison to the size of the tool which creates the layer. The size of the tool means anode size in this process, laser spot size in laser powder bed fusion (LPBF), electron beam diameter in electron beam melting, liquid drop size in ink jet printing and extruded path diameter in fused deposition modelling.

What if the layer size is equal to the tool size, then there will be no need for scanning, there will be no more problem for finding the right overlap between adjacent scanned lines, there will be no need to be careful to integrate each and every bit of such a big layer, there will also be no need to search right parameters to join such big layers. Thus, fabricating with a layer size equal to the tool size gives advantages over fabricating with a big layer size. It does not mean that such small layer size is recommended, it also does not mean that future of layerwise fabrication is with such small layer size. It only means that there is advantage with such small size which is not available with such big size – is there any method to be benefitted from such advantage. In a usual layer-tool setting, the layer by virtue of being a layer is big while the tool by virtue of being a tool is small – these sizes are two extremes. By selecting a basic building block smaller than and other than a layer will provide a compromise (a middle path) between two extremes.

A layer has some finite thickness and is similar to a rectangular plate which can be considered as an assembly of many smaller cubes or many smaller volumes similar to a shape something similar to a cube. These cubes make a layer and then make a 3D model. These cubes may not make a layer and then can make a 3D model. A cube may look like a layer and a layer may look like a cube. But, the basic difference between a cube as a basic building block and a layer as a basic building block is a build direction. An assembly of layers has only one fixed build direction that is perpendicular to all layers since an assembly of layers can have only one normal passing through all layers. An assembly of cubes has only one fixed build direction that is perpendicular to all cubes, if a cube is a layer. An assembly of cubes can have many build directions. Since a cube has six faces, it can have a maximum of six directions moving away from six faces. Since fabrication is going to happen on a substrate or a platform, then there cannot be any build direction towards the substrate resulting in a maximum of five build directions. It is not usual but also not impossible to have a case of six build directions where fabrication will take place without using a substrate or platform – such as in two-photon polymerization where fabrication happens in the middle of liquid and can be proceeded in any direction.

A 3D model, consisting of such cubes or cube type elements also referred as voxel, can have many build directions, if algorithm is well developed, and will not be affected by the limitations of layerwise fabrication. Using voxels instead of layers in an example of pillar and arm, there will be no more need for orienting the cathodic plate or using more than a single CAD file to overcome the limitations of layerwise fabrication.

9.2.2 Colloidal Solution Based Additive Manufacturing

Deposition of metal ions or using salt solutions does not allow many types of materials to be deposited. A colloidal solution provides versatility in materials choice; materials of any type such as polymers, alloys, ceramics are available in the form of colloids. These unlike metal ions can be of either charge depending on the types of additives and solutions and can be deposited on either electrode termed as either cathodic electrophoretic deposition or anodic electrophoretic deposition (Pikalova and Kalinina 2019). Voltage is applied across electrodes to drive these colloidal charged particles to electrodes, while in case of ionic solution based AM, voltage is applied to ionize the solution and drive the ions. Thus, in this AM process, solvent is chosen such that it will not ionize readily; if it will ionize then ions will change pH of the solution and thus disturbs the stability of colloids or ions will instead be deposited and will interfere with the deposition of charged colloidal particles. The particles need to be smaller than one micron so that their positions in the solution will not be affected by gravity. If particles are bigger then they will either sediment on the bottom of the solution or they will be in the process of sedimenting. If they will be sedimenting, then the solution will have fewer particles on the upper side of the solution and more particles on the bottom side of the solution – it will result in a thinner deposition on the top and a thicker deposition on the bottom of a vertical electrode. Thus, bigger particles will not provide a deposition of uniform thickness but a deposition of graded thickness. Since in this process, fabrication happens by the addition of small particles by small particles, the process is slow and suitable to make small parts.

Conductivity of an electrode is important for uniform deposition to occur. If the conductivity is low then there will be poor deposition on the electrode and vice versa. This fact can be utilized to accomplish the deposition at a particular area on an electrode by increasing the conductivity of that particular area of a low conductive electrode. Conductivity of a photoconductive electrode can be increased by irradiating with light. Thus, if a photoconductive electrode in the form of a plate is used as an electrode, then it can be irradiated with light on its back side to create deposition on its front side; the front side is in contact with colloidal solution. If the irradiation of light makes a pattern on the back side of the electrode, then the deposition of charged particles makes a pattern on the front side – thus, a 3D structure though of limited geometry on an electrode can be created (Mora et al. 2018).

References

ASTM F2792-12a (2012) Standard terminology for additive manufacturing technologies (withdrawn 2015). ASTM International, West Conschohocken

Brant A, Sundaram M (2016) A novel electrochemical micro additive manufacturing method of overhanging metal parts without reliance on support structures. Procedia Manuf 5:928–943

Castellano PMH, Vega ANB, Padilla ND et al (2017) Design and manufacture of structured surfaces by electroforming. Procedia Manuf 13:402–409

Coupek D, Friedrich J, Battran D, Riedel O (2018) Reduction of support structures and building time by optimized path planning algorithms in multi-axis additive manufacturing. Procedia CIRP 67:221–226

Goh GL, Agarwala S, Tan YJ, Yeong WY (2018) A low cost and flexible carbon nanotube pH sensor fabricated using aerosol jet technology for live cell applications. Sensors Actuators B Chem 260:227–235

Habib MA, Gan SW, Rahman M (2009) Fabrication of complex shape electrodes by localized electrochemical deposition. J Mater Process Technol 209(9):4453–4458
Kamraj A, Lewis S, Sundaram M (2016) Numerical study of localized electrochemical deposition for micro electrochemical additive manufacturing. Procedia CIRP 42:788–792
Lin JC, Chang TK, Yang JH et al (2010) Localized electrochemical deposition of micrometer copper columns by pulse plating. Electrochim Acta 55(6):888–1894
Manukyan N, Kamaraj A, Sundaram M (2019) Localized electrochemical deposition using ultrahigh frequency pulsed power. Procedia Manuf 34:197–204
Matsuzaki R, Kanatani T, Todoroki A (2019) Multi-material additive manufacturing of polymers and metals using fused filament fabrication and electroforming. Addit Manuf 29:100812
Mora J, Dudoff JK, Moran BD et al (2018) Projection based light-directed electrophoretic deposition for additive manufacturing. Addit Manuf 22:330–333
Paul R, Anand S (2015) Optimization of layered manufacturing process for reducing form errors with minimal support structures. J Manuf Syst 36:231–243
Pikalova EY, Kalinina EG (2019) Electrophoretic deposition in the solid oxide fuel cell technology: fundamentals and recent advances. Renew Sustain Energy Rev 116:109440
Wilkinson NJ, Smith MAA, Kay RW et al (2019) A review of aerosol jet printing – a non-traditional hybrid process for micro-manufacturing. Int J Adv Manuf Technol:1–21

Chapter 10
Additive Non-layer Manufacturing

Abstract The progress of additive manufacturing (AM) is hindered due to some of its unavoidable demerits. The cause of one of its demerits, that is staircase effect, is due to a fixed build direction. Additive non-layer manufacturing (ANLM) processes such as CLIP, 2PP and CNC accumulation do not have a fixed build direction, and these processes are therefore free from such demerit. These processes thus show a promising route to fabricate an AM part free from any inaccuracy arising due to staircase effect. This chapter describes disadvantages of additive layer manufacturing processes and analyzes various ANLM processes.

Keywords Non-layer · Two photon · CNC accumulation · Photopolymer · Staircase effect · Repair

10.1 Introduction

There are few additive non-layer manufacturing (ANLM) processes such as layerless fused deposition modeling, CNC accumulation, continuous liquid interface production (CLIP) and two-photon polymerization (2PP). Layerless fused deposition modeling comes under solid deposition process while the remaining three processes come under photopolymer based process as shown in classification given in Fig. 10.1. These processes are emerging and have not found acceptance and applications as additive layer manufacturing (ALM) processes have. However, these processes provide a different methodology to fabricate a part which could be of interest if this methodology helps emerge new AM processes applicable to make high-value ceramic or metallic parts. Since ANLM processes are free from layers they are also free from the demerits that come along with layers. This chapter provides various disadvantages of layer upon layer process and describes ANLM processes to check if these processes have potential to demonstrate alternative path to overcome the demerits of ALM processes.

S. Kumar, *Additive Manufacturing Processes*,
https://doi.org/10.1007/978-3-030-45089-2_10

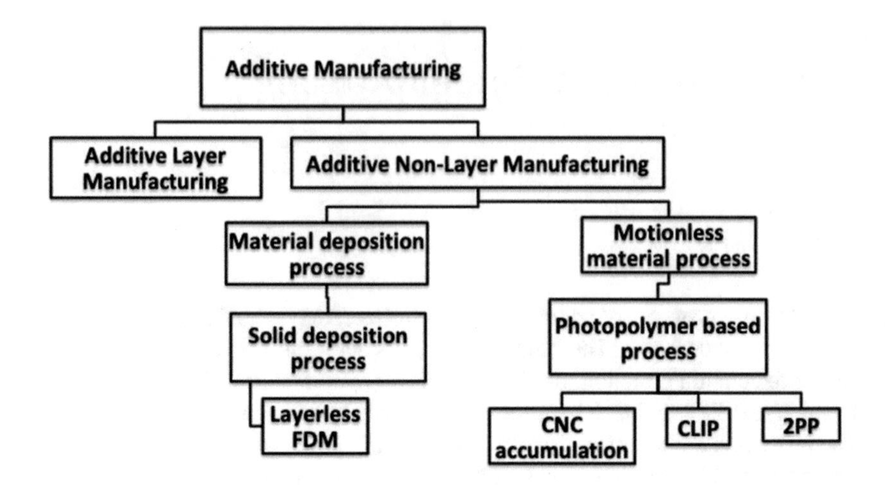

Fig. 10.1 Classification of ANLM

10.2 Disadvantages of Additive Layer Manufacturing

Conversion of a CAD model into layers in Additive Layer Manufacturing (ALM) has a well-known advantage – the problem of making a complex 3D part becomes a problem of making complex 2D layers. Since the fabrication of a complex 2D layer is easier than the fabrication of a complex 3D part, ALM replaces the difficult fabrication process by a simpler fabrication process. However, ALM has certain disadvantages such as staircase effect (Schmidt et al. 2017), need for support structures and problem in repair of an AM part. These are given below.

10.2.1 Staircase Effect

During the fabrication of a curved part, layers do not exactly coincide with the periphery of a curve which causes gaps between layers and the curve–these gaps are inherent deficiency of ALM and can never be eliminated because a straight line will never coincide with a circular curve. The gap can be minimized by decreasing the layer thickness, but there is limitation by which the thickness can be decreased. If a planar layer is replaced by a curved layer, then gaps can be eliminated in some geometries, but there is no AM process which is developed for curved layers, though there has been attempts with extrusion based process and curved laminates either to make parts or to make features on an existing part (McCaw and Urquizo 2018). Staircase effect is mentioned in Chap. 3 as well.

The causes of staircase effect or stairstepping effect are the following: (1) the build direction is vertical and (2) the progress of feature makes an angle with the vertical direction or there is a vertical curved feature. For example, if a right cylinder is made with its base on the platform, there will not be any staircase effect because the build direction does not make an angle with the centre line of the cylinder or the centre line is not tilted from the build direction (Fig. 10.2a). The cylinder is made up of horizontal planar layers with circular peripheries (disc); these discs are vertically aligned constituting a cylinder, there is no gap anywhere or no staircase effect. If the same cylinder is fabricated in a tilted position, horizontal planar layers or discs need to be made to constitute this tilted cylinder, since discs grow in a vertical direction they have vertical straight boundaries; these vertical straight boundaries do not conform the oblique boundary of the tilted cylinder; the generation of gap between vertical and oblique boundaries is staircase effect (Fig. 10.2b, c). If the discs will have oblique boundaries, then there will not be any staircase effect because then there will not be any generation of gap, if the build direction will be oblique then the disc will have oblique boundaries. If the build direction is not fixed but tilts with a tilt in a feature, there will never be any such staircase effect. If a process has variable build directions, this will be an antidote to the staircase effect.

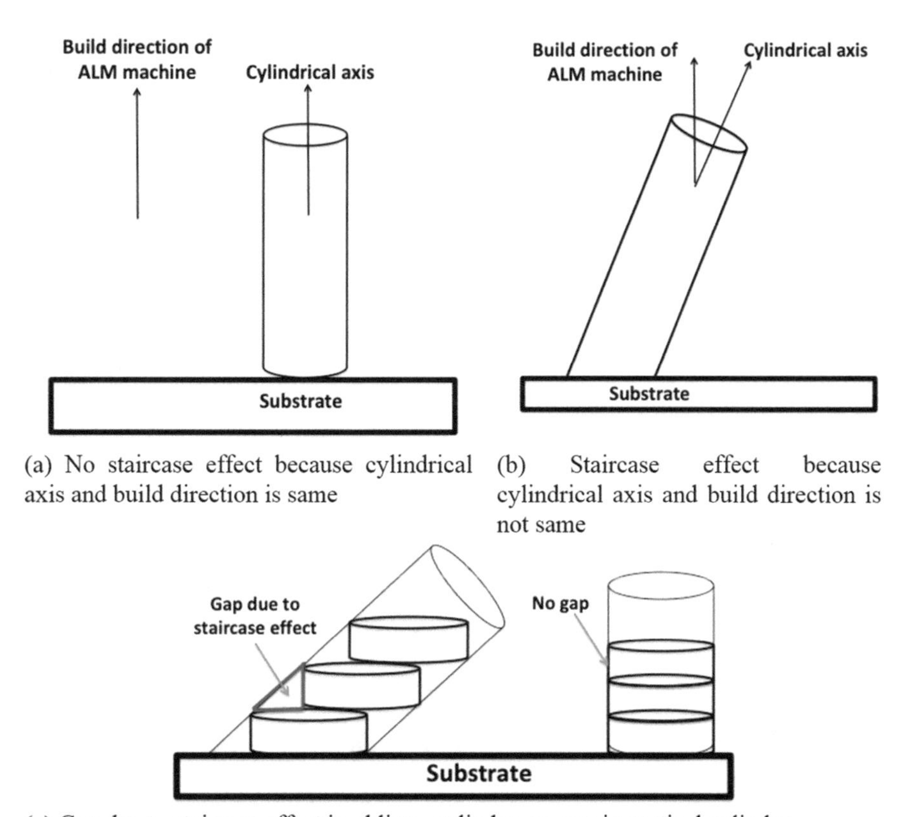

(a) No staircase effect because cylindrical axis and build direction is same

(b) Staircase effect because cylindrical axis and build direction is not same

(c) Gap due to staircase effect in oblique cylinder, no gap in vertical cylinder

Fig. 10.2 Cause of staircase effects: (a) an example of no staircase effect, (b) an example of staircase effect, (c) gap due to staircase effect

This effect arises because of position of one layer on another layer. In order for this effect to happen, more than one layer needs to be involved. This implies that if a layer is perfectly made, geometries within the layer are perfectly demarcated and filled up; these will have no bearing on the staircase effect.

10.2.2 Need for Support Structure

A tilted structure can be self-supporting in AM but if a tilt exceeds a certain angle depending on the weight and geometry, a support structure is required to maintain ongoing fabrication and to counteract potential collapse due to gravity (Mezzadri et al. 2018). The support structure increases the time of fabrication and post-processing and requires extra material and cost (Paul and Anand 2015). It is given in Chap. 12 as well.

10.2.3 Problem in Repair of an AM Part

If a feature of a part made in an AM machine is broken, then in order to repair it, layer by layer fabrication for that part needs to be revisited (Wilson et al. 2014). If the broken feature is confined within the last few layers, the part is machined or the material is removed layer by layer and is continued so that the broken feature is completely removed layer by layer; the removal or machining of the broken feature may not be possible unless other nearby features are simultaneously removed. If the removal or machining of a broken feature may be possible without removing other nearby features, then deposition of materials or coating of materials may not be possible without getting blocked or hindered by nearby features. This may warrant removal of not only broken features but also nearby features. The nearby features do not need repair but they are also getting disturbed and removed in an effort to repair broken features. Depending upon the geometry, there might be several unbroken features that need to be sacrificed in order to repair a single broken feature. Several unbroken features mean many unbroken features, but do not mean all unbroken features. It does not mean that there will never be cases when several unbroken features will not mean all unbroken features. For example, considering a case of spur gear that was fabricated in an AM machine, the build direction will be parallel to the edge of the teeth of the gear; this implies that the build direction is the same as the axis of rotation of the gear. In this case, if a tooth is broken, since the tooth extends from the first layer to the last layer of the part, machining to remove the broken tooth will cause removal of all layers. Removal of one tooth means removal of all teeth. In this case, repair is not possible as there is no machined part remained to be repaired upon, any attempt more to repair the part will be akin to replacement of the part. It does not mean that there is no hope left if anyhow a tooth of the gear gets damaged other than just replacing the gear. There are methods available in AM as well, such as repairing by changing the orientation after capturing the image of the damaged tooth by reverse engineering (Anwer and Mathieu 2016). But, the above method was shown to demonstrate that the benefit of AM achieved in repair of a complex part is no longer achievable if the damage is no longer confinable within the last few layers. In other words, the above method has demonstrated that damage done at a certain location of a part is not repairable because the attempt was going to be done to repair layer by layer. Since manufacturing in AM is done layer by layer to make a complex part, it might not be possible that the manufacturing in the realm of repair in AM will be different than layer by layer and will still be able to repair a severely damaged complex part. The above method has demonstrated that there is a basic flaw in additive layer manufacturing (ALM) – the flaw was exposed when the attempt was done to repair. (How to repair a part is given in Chaps. 3 and 6.)

In an ideal case, a repair needs to be confined to the damage site so that the repair can be accomplished without damaging and repairing larger area than the area of the damage site. It is possible if an AM process comes which does not see manufacturing through the prism of a layer and secondly has the ability to have direct access to the problem site so that it could try to repair. In the present ALM, a tool (energy beam, nozzle) is exposed only to the upper side of a layer; the tool has no access to either lower side or edge (side surface) of the same layer. There is no need for the tool to be concerned with other sides other than just upper side because manufacturing and growth happen only through the upper side. ALM is happening only through the upper side of the layer and it will not be an exaggeration if ALM will instead be called 'upper side ALM'. Since there is no need for the tool to work on other sides, the tool does not work on other sides, what if there is a need – a damaged part is kept on a platform and there are damages at various locations of the part, and damages could be repaired if the tool can manoeuvre and access them. Even if there is a need as such, the tool cannot directly approach the damaged sites, tool will only move through the layers; the process is handicapped to move up through layers, the process does not give freedom to the tool to move arbitrarily even if the need arises; if layer-by-layer phenomenon has given advantages, these advantages have come at a cost. There are several AM machines available which are equipped with tools fitted in multi-axis machines or multi-axis robots, but this is the capability of a machine (Tsao et al. 2018) and not the capability of a process; the tool equipped in such machines is still not free from limitations imposed by the process when the machine is using the tool to perform the process.

10.3 Additive Non-layer Manufacturing Process

ANLM processes are layerless fused deposition modelling (Kanada 2015), CNC accumulation (Chen et al. 2011), continuous liquid interface production (CLIP) (Janusziewicz et al. 2016) and two-photon polymerization (2PP) (Wu et al. 2006).

There is some research on low melting point metals to develop a free form structure without any support structures (Rangesh and O'Neill 2012).

ANLM processes are described below.

10.3.1 Layerless Fused Deposition Modeling

Fused deposition modeling (FDM) is an extrusion based process where materials are extruded from a nozzle or printhead and deposited line by line to make a layer. During the formation of a layer, the height of the nozzle from the layer does not change, the nozzle moves up after the last line of the layer is deposited; the nozzle gets opportunity to move up only after the completion of each layer. This is a usual FDM which is widely used and known. What if the nozzle does not wait for the completion of a layer before it moves up. In layerless FDM, the nozzle constantly moves up, while it is depositing a line, simultaneously moving up and moving forward is making the deposition path spiral (O'Dowd et al. 2015). Moving up in such a way does not let a layer to be made; if there are no layers, there does not exist a problem to join them, if there is no inter-layer joining, there will not be any issue with inter-layer debonding. Constantly upward moving nozzle means that there will always be an empty space; this process does not have the capability, in principle, to fill up the empty space that has been made by a spiral path. For example, for making a thin cylinder by spiral path, there will always be space at the centre to accommodate a thin wire. A solid thin cylinder can be made if the extruded material has low viscosity, so that the material will move inwards and fills up the gap at the centre, but if the material has low viscosity, it will also move outwards increasing inaccuracy. A solid thin cylinder will thus be made but not due to the merit of the process but because of the demerit of the process. For making a big solid cylinder, the nozzle has again to go down and start moving up again, the movement of the nozzle will be obstructed by the structure it has already made; the process is thus not able to make a solid part and is able to make only hollow parts. Additive layer manufacturing might be having many disadvantages but it at least provides a route to make a solid part, while this process might be having many advantages but it does not provide any route to make a solid part (Fig. 10.3).

If the process makes only hollow objects such as an empty cylinder, the process does not distinguish itself from FDM which already makes such cylinder by myriad machines. Layerless FDM distinguishes by making empty cylinders with increasing or decreasing diameter resulting in conical or inverted conical type structures, without taking help of support structures; the process has higher scope than usual FDM when there is a need to change diameters and vertical angles of such conical structures. In FDM, when such structures are made, deposited line is partially supported by previously deposited line, depending upon a change in diameter, the next deposited line will either slip and fall or break though and fall, which can be prevented by creating supporting structures. In layerless FDM, there is continuous deposited spiral line which provides lateral support and prevents the structure from falling. This

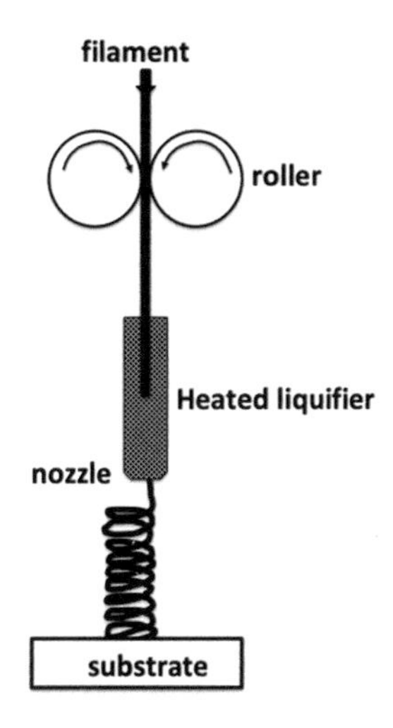

Fig. 10.3 Schematic diagram of layerless FDM

process thus demonstrates a method to avoid a support structure, which is one of the disadvantages of ALM (Kanada 2015).

10.3.2 CNC Accumulation

This is a photopolymer based process in which a substrate is immersed in photopolymer liquid and a part is fabricated inside the liquid (Chen et al. 2011). In order to start solidifying or curing liquid from the substrate, light beam needs to reach to the substrate without solidifying or curing the liquid on the way; it is possible if the beam is guided to the substrate through an optical fibre. The beam coming from the fibre will solidify the liquid between a fibre tip and the substrate, the solidified material will attach on the substrate and the fibre tip, Teflon film on the tip will decrease the adhesion between the solidified material and the tip causing the material to attach solely on the substrate. An increment in the attached material happens as per the requirement rather than as per the pre-set layer by layer values, thus making this process additive non-layer manufacturing (ANLM) (Fig. 10.4).

The attached material grows in the direction of fibre tip, when the tip moves, the solidified material will follow it, controlling this movement will create a desired structure. In order to move the tip, it needs to be stiff, the tip is supported inside a plastic rigid body and is called a tool. Movement of this tool creates a structure, makes a part and repairs parts; this movement of this tool is different from that movement of tools in ALM. This movement is not confined to a layer or a plane while creating structures, that movement is confined to a layer while creating structures. This movement can furnish many build directions, while that movement furnishes a single fixed build direction. Since this movement furnishes many build directions, this process is relatively free from staircase effect. This movement

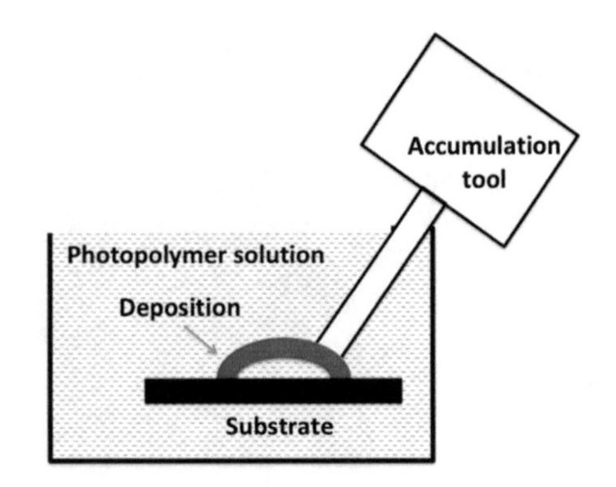

Fig. 10.4 Schematic diagram of CNC accumulation

creates structures on many sides of a substrate, at least three sides of a rectangular substrate, while that movement creates structures only on an upper side of a substrate. The ability of this movement to create structures on many sides of a substrate gives the process an ability to repair a part without changing the orientation, to repair a part having many damages on many sides without changing the orientation. This ability of this process distinguishes it from other AM processes (Simonelli et al. 2014) which are distinguished not to have this ability, whose generic fundamental demerit is exposed when this ability is tested.

The movement of the tool is facilitated by attaching it on a CNC machine; the movement adds materials on a substrate (while the movement of a cutting tool in a usual CNC machining removes materials from a block); a tool comprising of large beam diameter and high laser power will add more material and speed up the fabrication while a tool comprising of small beam diameter and low laser power will add small material causing an increase in resolution (Pan et al. 2014). A combination of tools will enable the fabrication of various geometries.

10.3.3 Continuous Liquid Interface Production (CLIP)

In inverse stereolithography, where a part is made upside down, a film is attached on inside of an exposure window so that a solidified layer can be easily detached from the window before resin is flown for the next layer to be formed. In CLIP, oxygen permeable window instead of a film is used, oxygen gas entering through the window reacts with free radicals and neutralizes them so that there will not be any more polymerization on or near the window. In absence of polymerization near the window, there will not be any solidified material on the window to be detached from; this makes the process free from problems of detachment, problems of repositioning the substrate after detachment, problems of breakage of tiny features during detachment and problems of not making large area because of the problem of detachment. Utilization of oxygen molecule to inhibit photopolymerization makes this process different from stereolithography.

CLIP is a continuous process; it is not a layer by layer process, there is no recurring phenomenon of layer formation and there will be no such recurring detachment. In CLIP, in order to make a part continuously, without being interrupted layer by layer, photopolymer through window is continuously exposed by beams while the substrate is continuously pulled up (Januszwiewicz et al. 2016). Continuous pulling up the substrate ensures that the photopolymer continuously flows in between the window and the substrate to be continuously cured or solidified by beams, fast pulling up will not let complete curing which means there will be non-uniformity or porosity in a part; slow pulling up will cause overcuring, making the part brittle, and will decrease fabrication speed (Fig. 10.5).

Since the process does not proceed layer by layer, the process saves time that could have been spent during inter-layer processing, that is the process saves processing time after completion of one layer and before starting the next layer, it increases the fabrication speed. Since there is no inter-layer joining, there are no seams between layers which could influence mechanical properties or change isotropicity.

Though, the process does not proceed layer by layer, it is still not free from layer by layer concept, the exposure on photopolymer or on window by beam makes 2D

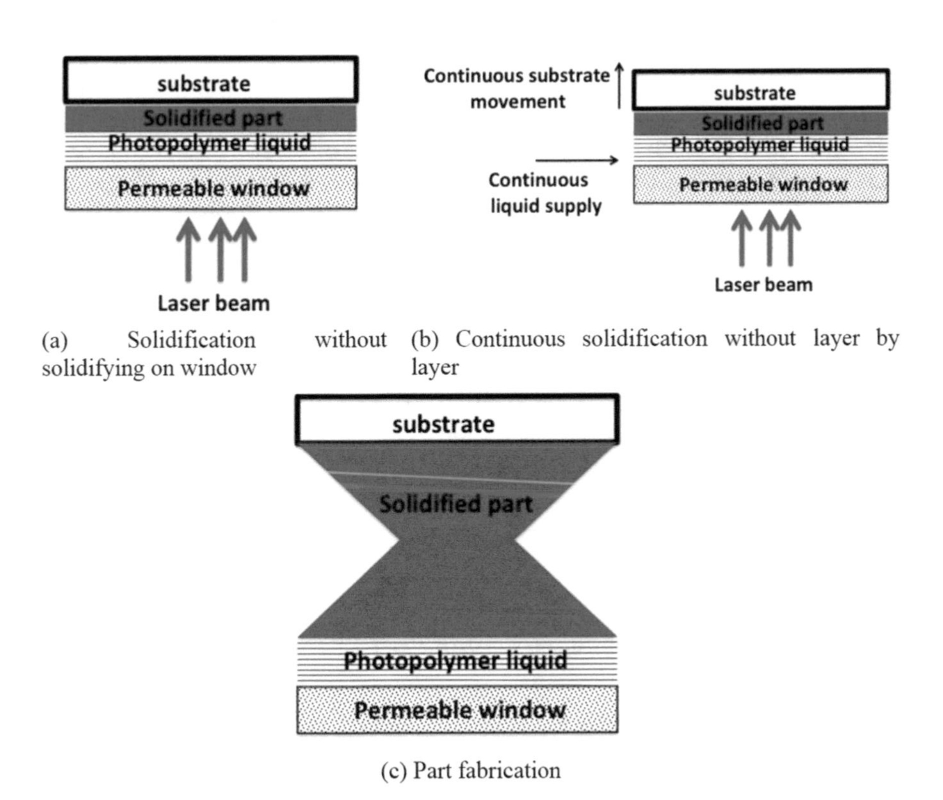

(a) Solidification without solidifying on window

(b) Continuous solidification without layer by layer

(c) Part fabrication

Fig. 10.5 Schematic diagram of CLIP

contour of a part in the same way as happens in a layer by layer process. For example, if a solid cone is to be made, the exposure on window will be of circular shape in both processes (CLIP and a layer by layer). After some duration of exposure, the exposure by beam will no longer be of the same size of the circle; it will change to fit the contour of the cone; when it will change, it will again be same to some nth layer of the layer by layer process. In layer by layer process, geometry of exposure by beam changes after a progress of a minimum of one layer thickness; the size of layer thickness is determined by the machine. In CLIP, geometry of exposure by beam can change very early or very fast because it is not related to an actual layer thickness a machine can allow, it is not limited by a machine. Since build direction does not change in both processes, there should be staircase effect in both processes, if staircase effect is not visible in CLIP, it is due to minimization of equivalent layer thickness in CLIP. This type of minimum layer thickness cannot be set in an actual layer by layer machine because of the limitation or tolerance of a machine. CLIP thus demonstrates how the limitation of an ALM machine can be overcome. This also shows that staircase effect is though a limitation of a process but this limitation is pronounced and visible not due to the limitation of the process itself but due to the limitation of the machine.

In a layer by layer process, if a 3D solid cone is fabricated, then the first layer will be made by scanning or exposing the photopolymer by a beam in a circular area; the effect of the scanning is the formation of a solid cylinder having height equal to the layer thickness, this cylinder instead of a cone is the natural outcome of a layer by layer process. Though, the aim is to make a cone but cone is not made, cone is actually approximated by making a number of cylinders. What if cone upon cone is made to make a cone. The process does not make a cone because the process does not see a cone during processing, the process only sees a circle and is not allowed to see the extension of a circle in 3D. CLIP is not different, it does not try to look beyond the plane though it is a continuous process, the beam can look beyond the plane, it can penetrate through the plane and solidify as per Beer-Lambert exponential law of absorption (Jacobs 1992). But, CLIP does not use the beam to go along the contour of a cone or to penetrate obliquely to make a cone directly rather than indirectly by approximation; if CLIP uses beam penetration as per the contour of a part extended in 3D, the process will in principle remove staircase effect. CLIP does not use the beam differently than how it is used in other layer by layer process. It is no exaggeration if CLIP is considered a very fast layer by layer process having no time to make a layer.

10.3.4 Two-Photon Polymerization (2PP)

The biggest demerit of additive layer manufacturing is that not a single part can be made without making a layer and making a layer would be one of the biggest problems if 100 nm resolution is sought. 2PP is a photopolymer based process which makes parts of size sub-100 micron having features with sub-100 nm resolution for

Fig. 10.6 Schematic diagram of 2PP

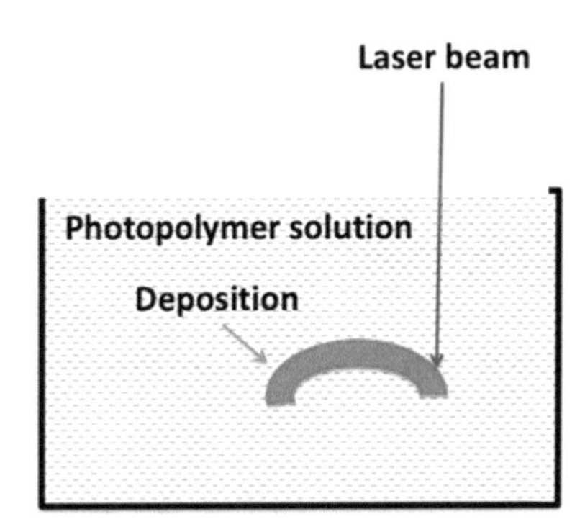

applications in various fields such as medical, optical and acoustical (Nguyen and Narayan 2017) (Fig. 10.6).

In 2PP, a laser beam in the form of a femtosecond pulse is required to excite the photopolymer and solidify it. In order to excite it, two photons within the interval of 10^{-15} s are required. One photon excites the molecule to an intermediate virtual state. If another photon is not available before the molecule loses its intermediate state and comes back to the ground state, the molecule will not able to reach the higher excited state. Thus, absorption of two photons present in space and time made available through a laser beam leads to excitation of polymers which causes polymers to form either free radicals or cations for polymerization. Energy contributed by two photons is necessary to excite a molecule in two-photon polymerization (2PP). If the same energy is contributed by a single photon, the molecule will not be excited because of the minor difference in quantum states in two cases (Fourkas 2016); therefore, 2PP is radically different from single photon polymerization or photopolymer bed process.

The probability to find two photons within such a short time period is possible only in an environment of high density of photons; such environment is found at the focal point of a laser beam. Therefore, when a laser beam irradiates the liquid, only the small volume of photopolymer present near the focal point of the beam gets cured while the vast amount of liquid facing the beam remains unchanged. Though, this tiny amount of solidification makes the process slow, and even a sub-mm size part is too big to be made by this process, the same tiny amount gives the process a capability to make small features and to have high resolution. The smallest feature made in a beam based process is limited by diffraction limit imposed by the beam. This process is able to make far smaller features than the wavelength of the laser used showing that the particle characteristic of a light is dominating over its wave characteristic in this process.

Tracing the beam in liquid will leave a trail of solidified material and this is how a 3D feature can be made by scanning the beam as per a geometry as per convenience (without going through layers). These features can be made on the surface of liquid or inside the liquid, but making inside the liquid will make the process free from availing feedstock (liquid) at reaction sites, features made inside the liquid remain self-supported and stable which neither sink nor buoy up.

References

Anwer N, Mathieu L (2016) From reverse engineering to shape engineering in mechanical design. CIRP Ann Manuf Technol 65:165–168

Chen Y, Zhou C, Lao J (2011) A layerless additive manufacturing process based on CNC accumulation. Rapid Prototyp J 17(3):218–227

Fourkas JT (2016) Fundamentals of two photon fabrication. 3D fabrication using two photon polymerization. Elsevier, pp 45–61

Jacobs FP (1992) Rapid prototyping and manufacturing: fundamentals of stereolithography. Society of Manufacturing Engineers, Dearborn

Janusziewicz R, Tumbleston JR, Quintanilla AL et al (2016) Layerless fabrication with continuous liquid interface production. PNAS 11(42):11703–11708

Kanada Y (2015) Support-less horizontal filament stacking by layer-less FDM. SFF Proc:56–70

McCaw JCS, Urquizo EC (2018) Curved-layer additive manufacturing of non-planar parametric lattice structures. Mater Des 160:949–963

Mezzadri F, Bouriakov V, Qian X (2018) Topology optimization of self-supporting support structures for additive manufacturing. Addit Manuf 21:666–682

Nguyen AK, Narayan RJ (2017) Two-photon polymerization for biological applications. Mater Today 20(6):314–322

O'Dowd P, Hoskins S, Geisow A, Walters P (2015) Modulated extrusion for textured 3D printing. NIP Digital Fabr Conf 1:173–178

Pan Y, Zhou C, Chen Y, Partanen J (2014) Multitool and multi-axis computer numerically controlled accumulation for fabricating conformal features on curved surfaces. J Manuf Sci Eng 136(031007):1–14

Paul R, Anand S (2015) Optimization of layered manufacturing process for reducing form errors with minimal support structures. J Manuf Syst 36:231–243

Rangesh A, O'Neill W (2012) The foundations of a new approach to additive manufacturing: characteristics of free space metal deposition. J Mater Process Technol 212(1):203–210

Simonelli M, Tse YY, Tuck C (2014) Effect of the build orientation on the mechanical properties and fracture modes of SLM Ti–6Al–4V. Mater Sci Eng A 616:1–11

Schmidt M, Merklein M, Bourell D et al (2017) Laser based additive manufacturing in industry and academia. CIRP Ann 66(2):561–583

Tsao C, Chang H, Liu M et al (2018) Freeform additive manufacturing by vari-directional vari-dimensional material deposition. Rapid Prototyp J 24(2):379–394

Wilson JM, Piya C, Shin YC et al (2014) Remanufacturing of turbine blades by laser direct deposition with its energy and environmental impact analysis. J Clean Prod 80:170–178

Wu S, Serbin J, Gu M (2006) Two-photon polymerization for three-dimensional micro-fabrication. J Photochem Photobiol A Chem 181:1–11

Chapter 11
Sheet Based Process

Abstract Sheet based process uses machining but is generally dealt in the realm of additive manufacturing (AM). It brings a question whether sheet based process is an AM and if it is not an AM then whether it is hybrid AM. It could be easy to know whether it is AM, but it could not be so easy to know whether it is hybrid AM because there exists no criteria and definition for hybrid AM. This chapter applies the concept of hybrid manufacturing to sheet based process to check whether it is hybrid AM. Various sheet based processes such as ultrasonic consolidation, laminated object manufacturing and friction stir AM are briefly explained.

Keywords Ultrasonic consolidation · Laminated object · Hybrid · Friction · Laser foil

11.1 Introduction

As per Oxford dictionary, a sheet is a large rectangular piece of cotton, fabric or paper. In the context of manufacturing, the word sheet is used for a rectangular piece of any materials such as metal, polymer, ceramic, paper. Sheet upon sheet fabrication gives the concept layer upon layer fabrication an opportunity to showcase a material that is exactly resembling to a layer the concept cannot ask for more – other materials or feedstocks such as powder, wire or hygdrogel have to go through processings and conversions before they reach the status of a layer which the sheet has already achieved. This exactness in resemblance gives some advantages – if a very big product needs to be fabricated then it could be better to start from the biggest available feedstock (sheet) rather than from a very small feedstock such as powder, when smaller feedstocks are not available then there will not be any alternative such as for making a paper product there are no paper powder or paper wire available but a paper sheet is available in abundance. Sheet upon sheet phenomenon also conveys a message – before another sheet is placed and joined there is an opportunity to place something between two sheets if joining does not damage this something – this something could be electronics (Varotsis

S. Kumar, *Additive Manufacturing Processes*,
https://doi.org/10.1007/978-3-030-45089-2_11

et al. 2018), sensor and fibre (Yang et al. 2009) and the final product could be a satellite or an automotive panel; if this something is not so exotic electronics or sensors but if this something is non-exotic ordinary powder then also there is an exotic opportunity to make a product having wear and corrosion resistance (Ho et al. 2020).

This chapter describes various sheet based processes and analyses whether these sheet based processes or layer upon layer processes are similar to other non-sheet based layer upon layer processes.

11.2 Ultrasonic Consolidation

Ultrasonic consolidation (UC), also known as ultrasonic additive manufacturing is, a layer upon layer process in which a 3D object is fabricated by joining metal foils layer upon layer using ultrasonic bonding and shaping using a milling machine (White 2003).

Ultrasonic bonding between foils is accomplished with the help of a solid metallic probe called sonotrode, which is pressed onto these foils. After pressing these foils, sonotrode is given ultrasonic energy so that the sonotrode vibrates with certain amplitude and frequency across foils (Fig. 11.1). A forced and vibrating sonotrode is enough to join foils. A vibrating sonotrode smoothens the interface between two foils, destroys oxide layer of the interface and exposes clean surface of one foil to the clean surface of the other foil, causes plastic flow at the interface of the foils – these result in the creation of a metallurgical bond between two foils. In order to extend this bond along lengths of the foils, sonotrode travels from one end of the foil to the other end of the foils while being pressed and vibrated; vibration happens perpendicular to the direction of the travel. Foils are joined layer upon layer so that the desired height could be reached. If sonotrode is moving in x direction, thickness of two foils is measured in z direction, then vibration is planned to occur in y direction (Fig. 11.1a). In order to prevent the foils from getting displaced or moved, these may be clamped. The combined action of force and movement does not increase the temperature of foils to their melting points, and thus the process is suitable for fabricating those parts which require to be fabricated at low temperatures, for example a part containing sensors or electronic circuits. Most important parameters for ultrasonic bonding are travel speed, vibration amplitude and applied normal force on sonotrode (Janaki Ram et al. 2007).

Milling machine is used to separate foils from a pool of foils (Fig. 11.1b), to trim foils at boundaries, to remove periodically excess material from bonded foils as per a CAD model, to make pockets on the ongoing build to accommodate fibres, antenna or sensors or electronic circuits, to smoothen the ongoing build for removing building defects and to control z-height (Friel and Harris 2013).

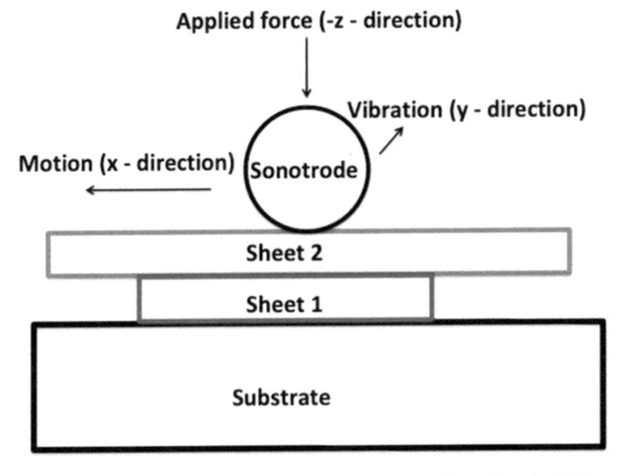

(a) Bonding by ultrasonic vibration in UC

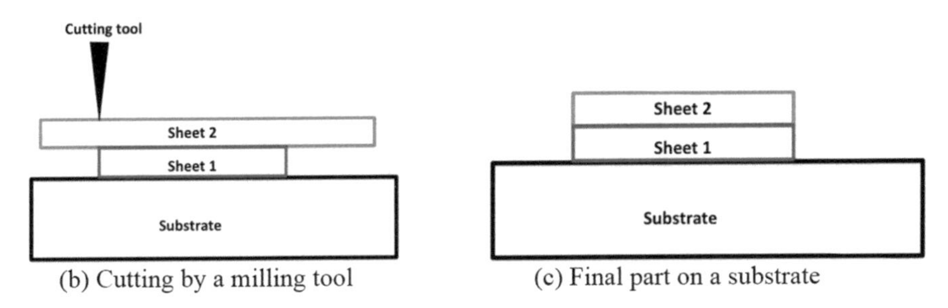

(b) Cutting by a milling tool (c) Final part on a substrate

Fig. 11.1 Part made in UC using sheet 1 and sheet 2: (**a**) sheet 2 is bonded on sheet 1 by moving sonotrode, (**b**) excess material of sheet 2 is removed by milling, (**c**) final part made on a substrate

11.3 Laminated Object Manufacturing

Laminated object manufacturing (LOM) is a sheet based layerwise process in which a 3D object is formed either by joining pre-cut sheets layer upon layer or by in-situ cutting and joining sheets layer upon layer. As the name of the process implies – it is about creating a laminated object or an object made up from several sheets (Wimpenny et al. 2003). It is a generic process, various types of sheets are used such as paper, plastic (Zhang and Wang 2017), metals (Hung et al. 2019), ceramics (Gomes et al. 2011), metallic glass (Li et al. 2017) sheets made from slurry (Liu et al. 2015) etc.; various types of joining methods are adopted, such as laser welding, glue, binder (Bhatt et al. 2019), freezing (Zhang et al. 2018), thermal pressing (Zhang and Wang 2017), cold pressing (Schindler and Roosen 2009) etc.; various types of cutting methods are adopted, such as using knife, milling (Schindler and Roosen 2009), laser cutting (Zhang et al. 2018) etc. In one variant, named as laser foil printing, metallic glass is processed by laser welding and laser cutting to make a part (Li et al. 2017).

Commercial LOM systems require continuous feeding of sheets to the platform for building an object sheet by sheet. Feeding is done by unfolding a roll which has several meters of sheet wrapped around it; unfolding the roll will enable fresh sheet to come over the platform where part of it will be separated from the sheet and be joined on the platform. The sheet will move forward so that a fresh part of the sheet will again come over the platform to be cut and joined on the platform – since the platform has already one sheet joined over it – the sheet will join on the sheet on the platform. This repetition of moving a fresh sheet and thereafter cutting and joining will give rise to sheet upon sheet or layer upon layer manufacturing process (Chiu et al. 2003).

Unfolding the roll to supply a fresh sheet requires used sheet to be folded on another roll on the other side of the platform. Thus, there are two rolls connected by a single sheet required to complete the feeding. Width of the sheet needs to be bigger than the width of the platform so that when part of the sheet is going to be cut, the cutting will not consume all width of the sheet or the cutting will not destroy the connection between two rolls; a continuity is required to ensure the feeding. This also means that for every fresh sheet there will be a used sheet – the process thus generates waste. The process thus cannot assertively claim to be better than machining in terms of generation of waste.

Cutting of the sheet happens by a knife or by a laser beam while joining may happen by pressing the sheet at a high temperature so that glue coated on the sheet will be activated and will subsequently join an underlying sheet. There are two types by which fabrication can be accomplished: (1) bond-then-cut (Liao et al. 2003) and (2) cut-then-bond (Thomas 1996).

In bond-then-cut type (Fig. 11.2), a complete fresh sheet is first joined or attached on a platform or on a previous sheet, then undesired part of the joined sheet is removed by cutting – this will leave remaining part of the bonded sheet to remain bonded and to become part of an end product. After joining, removing by cutting may not be easy especially if cutting is not at the boundary of a sheet. In order to facilitate removing by cutting, anti-glue liquid or powder is sprayed on selected area before joining so that the selected area is not strongly joined and can be removed

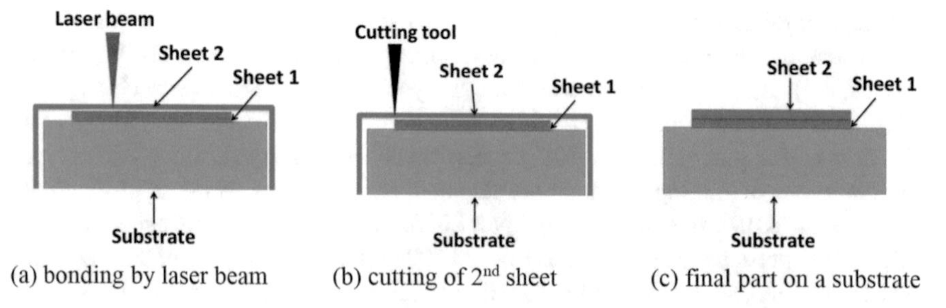

Fig. 11.2 Bond-then-cut type LOM: (**a**) sheet 2 is bonded by laser beam with sheet 1, which is attached on a substrate, (**b**) after bonding, excess material from sheet 2 is removed by a cutting tool, (**c**) final part on a substrate

without damaging underlying sheet. Spraying thus requires information from a CAD file and extra time, which increases fabrication rate. Another way to facilitate removing is to do cross-cutting or cross-hatching on a selected area several times so that to-be-cut area becomes weak and fragile, which can be removed afterwards easily; this type of removal is also known as decubing – this takes time and a lot of cutting energy, which makes the process energy inefficient.

In cut-then-bond type (Fig. 11.3), a sheet is cut or a pattern is made on a sheet as per the corresponding section of a CAD file, the sheet is not cut at the boundary or it is not cut in such a way that it will be severed from a continuous roll. If a sheet will be severed from the roll before the joining then, it will lose alignment or in worst case it will no longer be tensioned enough to be able to be cut. After cutting, the sheet is then joined on a platform or on previous sheet followed by cutting at the boundary and detaching the sheet from the roll. In this type, there is no hassle of cross-hatching or decubing – thus, this type can be preferred for simplicity and being energy efficient. But, the action of joining itself may cause misalignment – thus, this type is not preferred for accuracy. This type is better than previous type when small area needs to be bonded while previous type is better than this type when small area needs to be removed.

11.3.1 *Why Cut-Then-Bond Type Is a Subset of Bond-Then-Cut Type*

In cut-then-bond type, sheet is first cut and then joined to make a part in LOM while in bond-then-cut type, sheet is first joined and then cut to make a part in LOM. These two types seem to be different but it is questionable whether they are really different.

LOM system is aligned for bonding; the position of bonding with reference to platform is already fixed. It is the cutting which changes its position to create different patterns. Since the position of the bonding is already determined, it will not change if cutting happens either before or after the bonding. In a case when cutting precedes, then it is for some reason (simplicity) and in another case when cutting

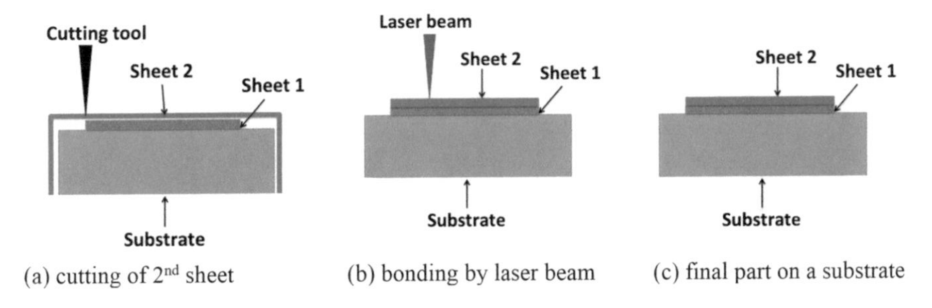

(a) cutting of 2nd sheet (b) bonding by laser beam (c) final part on a substrate

Fig. 11.3 Cut-then-bond type LOM: (**a**) excess material from sheet 2 is removed by a cutting tool, (**b**) after cutting, sheet 2 is bonded with sheet 1 attached on a substrate, (**c**) final part on a substrate

follows then it is for the sake of another reason (accuracy). In both cases, it is the bonding which is decided firstly – a cutting which precedes bonding does not change the fact that the cutting is going to be decided as per the pre-determined and pre-fixed position of bonding and not the vice versa. Therefore, cut-then-bond type is a special case of bond-then-cut type in which case bonding is delayed for the sake of convenience.

It does not imply that all cut-then-bond types are subsets of bond-then-cut type. Following example shows why the following cut-then-bond type is different than previous cut-then-bond type and is not a subset of bond-then-cut type.

A sheet is cut as per a CAD file so that it is ready to be bonded. It needs to be transported to a platform where it will be bonded. It needs to transported well so that its orientation should not change. If its orientation will change, it will be positioned and bonded differently then it will make a different part. If a sheet is complex, it requires to be aligned on a platform with the help of fixtures otherwise it will again bond differently, it does not have luxury as in a previous example to have pre-aligned transportation and bonding. If the sheet will be cut after bonding, then it will be free from such problem of alignment during transportation and final positioning. In this example of cut-then-bond type, bonding of a sheet is not determined by pre-fixed position of bonding, cutting is done independently. If cutting of a square ABCD is going to be done then no side (AB, BC, CD, AD) of the square is a preferred side, it is the orientation of ABCD that will decide which side will align with which side of the platform. While in a previous example, ABCD portion of a continuous sheet is fixed with respect to the platform (Fig. 11.4a, b), the position of AB and CD cannot be changed arbitrarily unless the position of the roll is changed by

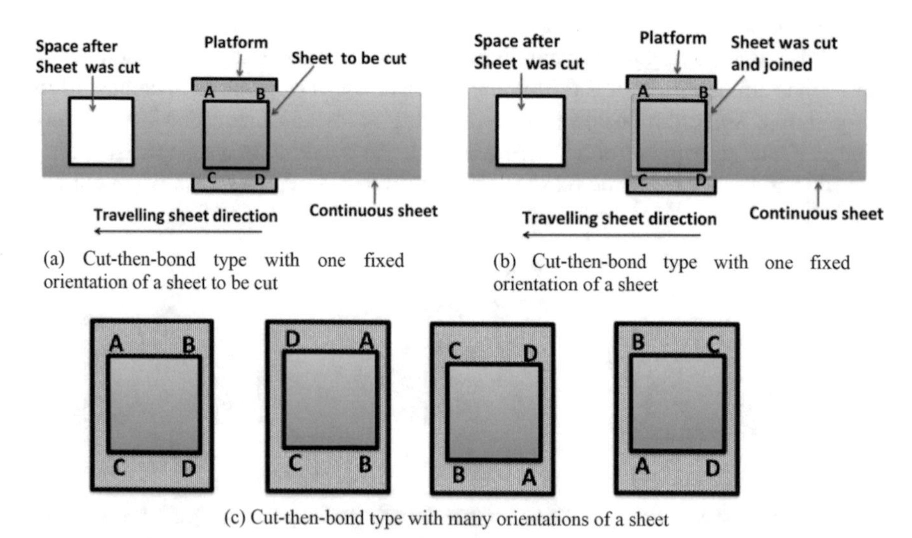

Fig. 11.4 Cut-then-bond type: (**a**) with one orientation of a sheet to be cut, (**b**) with one orientation of a sheet cut and joined (**c**) with many orientations of a sheet

disassembling LOM system. Therefore, cut-then-bond type in this example (Fig. 11.4c) is different from cut-then-bond type in the previous example (Fig. 11.4a). Thus, cut-then-bond type and bond-then-cut type in this example is drastically different from cut-then-bond type and bond-then-cut type in the previous example.

11.4 Why Sheet Based Process Is Not Additive Manufacturing

ASTM defines AM as 'a process of joining materials to make objects from 3D model data, usually layer upon layer, as opposed to subtractive manufacturing methodologies' (ASTM 2012). The definition implies that AM is opposed to subtractive manufacturing methodologies, and subtractive manufacturing does not play main role in AM. There are many processes which do not fulfil the definition, and as per the definition they do not come under AM but these are still mentioned as AM. Most prominent examples are sheet based process (SBP) such as ultrasonic consolidation, laminated object manufacturing, friction stir additive manufacturing, metal laminated tooling (Himmer et al. 2004), laser foil printing etc.

Sheet based processes use machining or subtractive manufacturing usually layer by layer; it implies that as joining happens layer upon layer, machining also happens layer by layer (Graff et al. 2010, Varotsis et al. 2018). For some geometries, machining might not occur strictly after fabrication or joining of each layer, and machining will occur after a certain number of layers as per the requirement, ease or complexity; but in all processes machining will never occur as a single step only in the beginning (as a pre-processing) or at the end (as a post-processing) – thus, machining is not an isolated step but is recurring essential intermediate steps to complete the process.

Without machining, not a simple part such as a cube can be made by these processes – for making a simple cube, sheet of the size of cross-section of the cube needs to be separated from the feedstock after joining of each layer; this is not possible without cutting or machining, and thus machining is an essential part of these processes. This gives rise to a question – whether machining is an essential part of these processes or joining is an essential part of these processes; in another way, which one is primary and which one is secondary, machining is primary or secondary.

In all AM processes, an action of adding or joining gives rise to a shape. In powder bed fusion (given in Chap. 3), it is the joining of the powders which determines the final structure; the place of action of joining on the powder bed guides rising of the structure, if there is no joining, no such structure and no such shape, there is no role of machining; absence of machining is not to declare that the process has reached to such level of perfection that it is beyond the need of machining, absence of machining is not even to partially support any claim or idea that the machining is going to be outdated, absence of machining is also not to assert that if occasionally in some cases machining is done as an intermediate step, then it will not improve the part; absence of machining is only informing what the very process is. If machining is done at a later stage, then it is to expose that structure prominently or to expose those actions of joining. Machining is not done to create a shape or machining is not approached to create an additional shape; recourse of machining will never happen if the action of joining is an action of well joining. Exclusion of machining at any stage is an ideal goal of AM and concurs with the definition of AM.

In other AM processes such as ink jetting, addition of ink determines the shape; in beam deposition, it is the joining of deposited materials which determines shape; in stereolithography, it is the joining of photopolymer which determines shape etc. It is not that shaping is the only thing expected in AM but it is the shaping which distinguishes AM from other manufacturing processes and it is pertinent to find what causes shaping in AM. In SBP, it is the machining which gives shape of a part; though without the aid of joining, the part will be disintegrated and there will not be any part more. But, it does not change the fact that the joining is not the cause of shaping in SBP – it makes this process not to have same cause of shaping as other AM processes do have. The simplistic outlook – since all AM process including sheet based process use adding and machining at some stage and all are same – does hide the fact that SBP is different from an AM process.

11.5 Why Sheet Based Process Is Considered Additive Manufacturing

There are following reasons why sheet based process (SBP) is considered AM:

1. ASTM mentions it as one of the seven categories of AM (ASTM 2012).
2. It is a matter of tradition; it has been called rapid prototyping (earlier version of AM) or additive fabrication for long (Paul and Baskaran 1996, Yan and Gu 1996). The process is more than two decades older than the definition of AM (Nakagawa 1979).
3. The process does addition layer upon layer (Obikawa et al. 1999).
4. There are examples where pre-fabricated plates (Bhatt et al. 2019) or pre-fabricated tapes (Schindler and Roosen 2009) are used and no such machining is done.

11.6 Why Sheet Based Process Is Not a Hybrid Additive Manufacturing

If SBP is not an AM process then it is relevant to find whether it will come under hybrid AM since it has both additive and subtractive components. There is no definition of hybrid AM, but there are some definitions for a process to come under hybrid manufacturing. These definitions are the combined action of two processes will be more than the net result of individual processes, that is 1 + 1 should be equal to 3 and not 2 (Schuh et al. 2009); the processes should act simultaneously and have a significant effect on process performance (Lauwers et al. 2014). In absence of any definition for hybrid AM, these definitions will act as criteria to find whether sheet based processes come under hybrid AM (Sealy et al. 2018).

11.6.1 An Ideal Hybrid Additive Manufacturing Process

In order to elaborate these criteria in the context of SBP, an example for hybrid AM related to sheet which fulfils these criteria is given in Fig. 11.5a. The figure shows a circular plate is cut from a large sheet by a laser beam, two such plates are then joined by laser welding to make a final part. Since cutting and welding do not occur at the same time but occur one after another, they do not satisfy the criteria of being simultaneous. In this example, when a laser beam cuts at the periphery of the plate, since plate is small, whole plate gets heated due to laser cutting; the plate continues to remain hot till the completion of joining by laser welding. Due to the plate being at high temperature, lower energy from a laser beam is required for laser welding; this saves some laser energy. Besides, welding by applying high laser energy creates high thermal gradient which causes bending of the plate due to stress – this will decrease the dimensional accuracy of the final structure.

In this case, the process laser cutting helps another process laser weld to perform better by decreasing the energy input requirement and removing thermal stress, that's why this combined process satisfies the condition of 1 + 1 = 3. In order to understand what would be 1 + 1 = 2, the process is delayed from the step of cutting to the step of welding; this delay results in the cooling of the plate, which does not decrease the requirement of energy input for welding, as well as properties due to welding are not equally good.

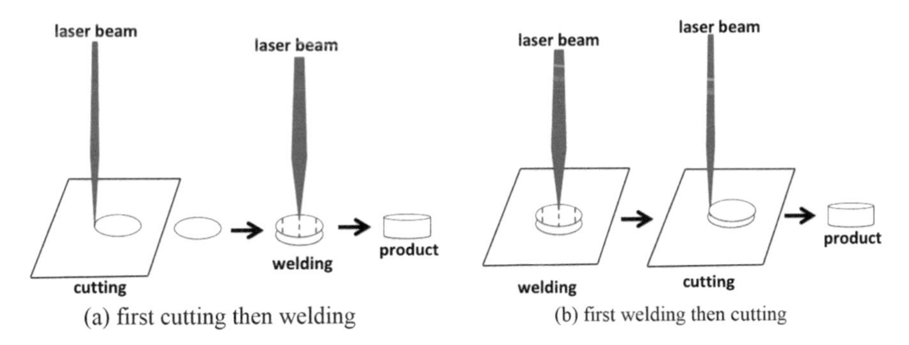

Fig. 11.5 Possible example of hybrid AM as per definition of hybrid manufacturing: (**a**) sheet metals are cut from the feedstock and then bonded, (**b**) sheet metal is bonded first and then cut from the feedstock

Considering a variation of above example (Fig. 11.5b) having a reverse sequence of operation – welding first and then cutting – where welding operation increases the temperature and allows the cutting operation to be faster, smoother and more energy efficient will be another example of hybrid AM.

11.6.2 Application of Hybrid Additive Manufacturing Criteria in Ultrasonic Consolidation

In ultrasonic consolidation (UC), joining of foils gives rise to anisotropic properties; it implies that the properties along the foil direction, along the transverse direction and along the z direction are different; it is due to anisotropicity in foil properties, defects in interfacial bonding, variation in grains at interface and adjacent zones and surface milling done on top surface (Gibson et al. 2010). After bonding or joining layers, when cutting is performed by a milling tool, the tool faces different cutting forces during milling different sides; in order to have smooth cutting, cutting speeds need to be optimized so that foils should not be peeled off. The action of bonding does not drastically increase the cutting performance, the action does not even help cutting to improve, and in case of peeling off the foil the action of cutting decreases the effect of action of bonding. Due to anisotropicity, even in absence of any bonding defects, cutting parameters need to be optimized to let the cutting be completed. In case of simple geometries, cutting parameters may not be optimized, but even in those cases the action of bonding does not decrease the cutting force and gives extra smooth surface. The action of bonding and cutting does not fulfil the condition of hybrid manufacturing, that is 1 + 1 = 3, as per the definition of hybrid manufacturing UC is thus not a hybrid AM.

It gives rise to a question if UC is neither AM nor hybrid AM then what it is. UC is also named as additive and selective subtractive manufacturing method (Kalvala et al. 2016). The definition of hybrid manufacturing given above is primarily written for machining processes where there are a number of combined processes which motivate to investigate how many of these combinations are more than just a combination, how many of these will be having a significant effect on the process performance, how many have a large influence on the processing characteristics (Lauwers et al. 2014) – in order to weed out those ordinary ones (just mere combination) from special ones. AM is not mature as these processes are, and AM is trying a number of possibilities to be accepted as a reliable manufacturing process; combinations of processes in case of AM are efforts to reach to that level of acceptance; in the course of efforts if AM exceeds that level of acceptance it would be an achievement, after all it is touted as one of the most disrupting processes; in some cases it is achieving but in majority of cases it is struggling to fulfil the basic expectation. The definition of hybridity or being in a hybrid state suggested or given earlier is only for general manufacturing, a comparatively mature process, and it is nowhere mentioned or stated that this definition is equally applicable to AM (a developing process).

In UC, attention is required so that the cutting will do the job of cutting and will not undo the job of bonding by peeling off the bonded foils; then it is not a question whether cutting and joining are making 1 + 1 = 3, the question is how to improve the process so that cutting and joining will not fail to make 1 + 1 = 2. Cutting or machining is incorporated in UC not to cause significant process improvement but is incorporated because there is no alternative of machining, it is not possible and has never been possible to make a small part from a bigger feedstock without making bigger feedstock smaller (given in Chap. 1). There can be some alternatives to machining: if machining is done by a high energy beam as used in laser cutting (Fig. 11.5), then the problem with cutting force can be avoided but it goes against the spirit of UC which is a solid state AM process and does not want to lose the advantage of being a solid state by using a high temperature process component which will disturb the status-quo. Another replacement of machining is to use pre-machined foils (Varotsis et al. 2018) so there will be not any need of intra-built machining and consequently there will not be any problem related with machining. Using pre-machined foils can give some advantage which has been given in Chap. 12. Either replacing the type of machining or replacing the need of machining will not let the UC to be the same UC; this is not a question about only a single process (UC) but all SBP; these processes do not come under hybrid AM as per the definition.

11.7 Friction Stir Additive Manufacturing

Friction stir additive manufacturing (FSAM) is an emerging AM process which utilizes friction stir welding (FSW) to join metallic sheets (Palanivel et al. 2015). In order to join two sheets, these are clamped together and FSW is performed on top of two sheets. For FSW to be executed, a pin (attached on a shoulder) is inserted at the edge of two sheets. It is inserted by rotating it in such a way that the pin crosses the interface between two sheets and partially enters the bottom sheet (Fig. 11.6). This position of the pin ensures that the change in the interface will be contributed by both sheets when the rotating pin moves from one edge to the other edge across the length. During rotation, heat will be generated due to friction between the pin and the sheets, and if material is ductile, the material will be displaced from its positionn, causing the interface to be filled up and be lost resulting in bonding of the sheets. Generation of heat by rotation and movement does not cause melting, therefore the process is called solid state process and is free from all defects that emanate from solid-liquid-solid transformation, such as hot cracking, porosity, loss of texture, anisotropic grain growths etc.

One movement or one pass of the tool or pin will not be enough to cover whole interface between two sheets if the width of the sheets are far bigger than the diameter of the pin, in that case a number of adjacent passes are required to bond two

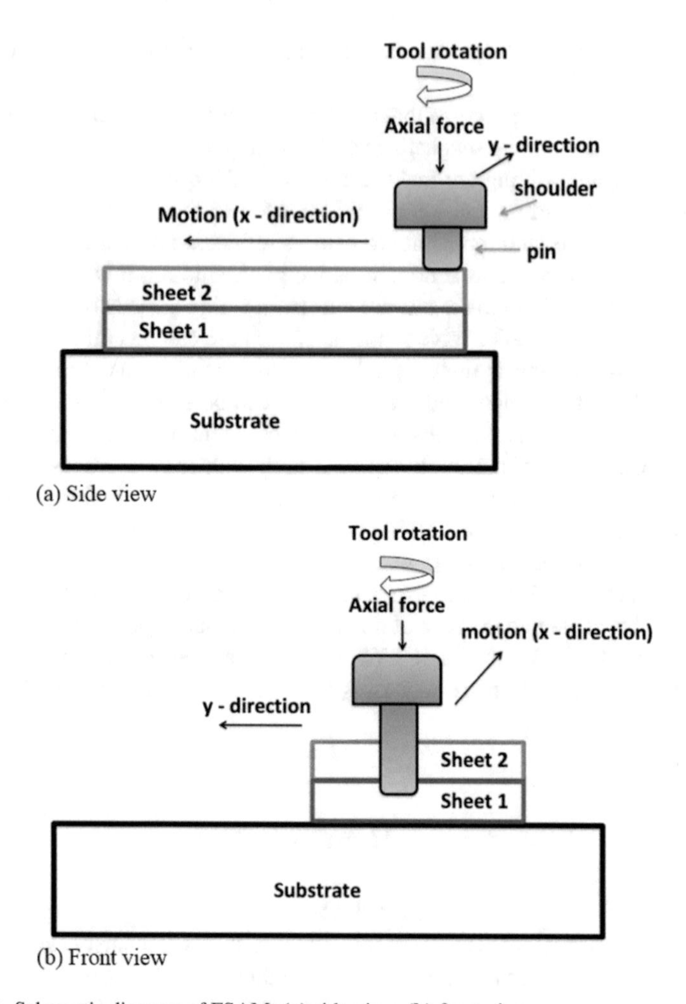

Fig. 11.6 Schematic diagram of FSAM: (**a**) side view, (**b**) front view

sheets or two layers. Depending on the required height, number of sheets will increase and they will be bonded by declamping the bonded sheet and then clamping with a new sheet on top of it; this will continue till the required height is reached. Figure 11.6a shows welding of two sheets (on a substrate) by a tool moving in *x*-direction while Fig. 11.6b shows its front view where pin of the tool is reaching the interface between both sheets.

The process has demonstrated the application of FSW to make bigger blocks from sheet metals which is in turn demonstrating FSW as an additional tool for adding or joining that has potential to be applied in AM. The process has not yet made a complex part. The hardness of stirred zone for most of the materials is more than the base metal (Zhao et al. 2019) but in some aluminium alloy, the hardness is reported to be lower than the base metal (Kalvala et al. 2016). The block shows higher ductility and strength than that of base metals (Palanivel et al. 2015).

11.8 Comparison Between FSAM and Powder Bed Fusion (PBF)

Since FSAM is not yet able to make a complex part, its potential will be clear if it can be compared with an established process such as PBF.

11.8.1 Surface Finish

In FSAM, tool causes depression on the surface which leads bulging at the end and non-uniformity on the surface; this requires machining to be done to trim the boundaries and mill the surface. In PBF, the surface does not require such machining to get appreciable surface finish.

11.8.2 Micro-features

Since tool applies pressure in FSAM, any feature that will not sustain this pressure will not be able to be formed. There is no such tool pressure in PBF. In FSAM, for making a small feature or joining on a small feature a small tool will be required while in PBF features of all size can be formed without any additional requirements. Using a number of tools for making a part consisting of multi-size features (in FSAM) is also against the basic ideal of AM that AM is a toolless manufacturing and the part fabrication should be free from size-specific tools.

11.8.3 Feedstock

Sheet metallic alloy should be ductile and should not be hard enough to break or make chips during stirring. The tool needs to be stronger than sheets and should have bigger size (height) than the thickness of the sheet. Powders do not have such mechanical property requirements.

11.8.4 Part Properties

FSAM does not give as anisotropic properties as fusion based PBF gives. But, FSAM furnishes inhomogeneous properties – top sheet gives higher hardness because it has not gone through annealing due to the repetitive motion of tools as other sheets have. Besides, depending upon the tool shape, the built consists of areas

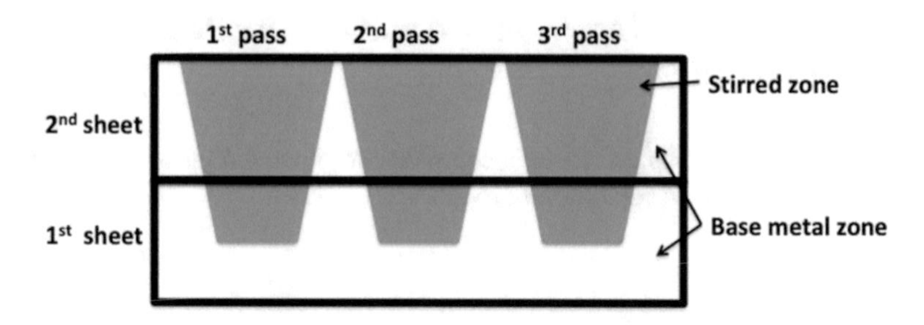

Fig. 11.7 Front view of two sheet metals bonded by three adjacent passes using truncated cone shape pin

of stirred zones and base metal zones giving rise different hardness. If tool is of truncated cone shape, then the stirred area of a sheet will constantly decrease across the thickness giving rise to more base metal zone on the bottom surface of the sheet. Figure 11.7 shows increasing gap between stirred zones made by adjacent passes.

11.8.5 Part Fabrication

While a design can be converted into a physical object automatically in a PBF system, attempt has not been done in FSAM to convert a design into part automatically; FSAM requires to incorporate machining to shape a part (Sames et al. 2016). If conventional machining such as milling is used for machining as is used in other friction based process such as UC and friction surfacing (Dilip et al. 2013), then it would be of interest to know whether FSAM will come under hybrid manufacturing as per the definition (Schuh et al. 2009). Since bonding gives inhomogeneous hardness which causes varied cutting force on a milling tool, the very bonding does not help cutting to improve drastically, and as per the definition FSAM will not come under hybrid AM.

References

ASTM F2792-12a (2012) Standard terminology for additive manufacturing technologies (withdrawn 2015). ASTM International, West Conschohocken

Bhatt PM, Kabir AM, Peralta M et al (2019) A robotic cell for performing sheet lamination-based additive manufacturing. Addit Manuf 27:278–289

Chiu YY, Liao YS, Hou CC (2003) Automatic fabrication for bridged laminated object manufacturing (LOM) process. J Mater Process Technol 140(1–3):179–184

Dilip JJS, Babu S, Rajan SV et al (2013) Use of friction surfacing for additive manufacturing. Mat Manuf Process 28:1–6

Friel RJ, Harris RA (2013) Ultrasonic additive manufacturing- a hybrid production process for novel functional products. Proc CIRP 6:35–40

Gibson I, Rosen DW, Stucker B (2010) Additive manufacturing technologies: rapid prototyping to direct digital manufacturing. Springer, New York

Gomes C, Travitzky N, Greil P et al (2011) Laminated object manufacturing of LZSA glass-ceramics. Rapid Prototyp J 17(6):424–428

Graff KF, Short M, Norfolk M (2010) Very high power ultrasonic additive manufacturing (VHP UAM) for advanced materials. SFF Symp Proc:82–89

Himmer T, Techel A, Nowotony S, Beyer E (2004) Metal laminated tooling: a quick and flexible tooling concept. SFF Symp Proc:304–311

Ho YH, Joshi SS, Wu TC et al (2020) In-vitro bio-corrosion behavior of friction stir additively manufactured AZ31B magnesium alloy-hydroxyapatite composites. Mater Sci Eng C 109:110632

Hung CH, Sutton A, Li Y et al (2019) Enhanced mechanical properties for 304L stainless steel parts fabricated by laser-foil-printing additive manufacturing. J Manuf Process 45:438–446

Janaki Ram G, Robinson C, Yang Y, Stucker B (2007) Use of ultrasonic consolidation for fabrication of multi-material structures. Rapid Prototyp J 13(4):226–235

Kalvala PR, Akram J, Mishra M (2016) Friction assisted solid state lap seam welding and additive manufacturing method. Def Technol 12:16–24

Lauwers B, Klocke F, Klink A et al (2014) Hybrid processes in manufacturing. CIRP Ann 63(2):561–583

Li Y, Shen Y, Chen C et al (2017) Building metallic glass structures on crystalline metal substrates by laser-foil-printing additive manufacturing. J Mater Process Technol 248:249–261

Liao YS, Chiu LC, Chiu YY (2003) A new approach of online waste removal process for laminated object manufacturing (LOM). J Mater Process Technol 140(1–3):136–140

Liu S, Ye F, Liu L, Liu Q (2015) Feasibility of preparing of silicon nitride ceramics components by aqueous tape casting in combination with laminated object manufacturing. Mater Des 66A:331–335

Nakagawa T (1979) Blanking tool by stacked bainite steel plates. Press Tech:93–101

Obikawa T, Yoshino M, Shinozuka J (1999) Sheet steel lamination for rapid manufacturing. J Mater Process Technol 89–90:171–176

Palanivel S, Sidhar H, Mishra RS (2015) Friction stir additive manufacturing: route to high structural performance. JOM 67(3):616–621

Paul BK, Baskaran S (1996) Issues in fabricating manufacturing tooling using powder-based additive freeform fabrication. J Mater Process Technol 61:168–172

Sames WJ, List FA, Pannala S et al (2016) The metallurgy and processing science of additive manufacturing. Int Mater Rev:1–46

Schindler K, Roosen A (2009) Manufacture of 3D structures by cold low pressure lamination of ceramic green tapes. J Eur Ceram Soc 29(5):899–904

Schuh G, Kreysa J, Orilski S (2009) Roadmap 'Hybride Produktion': Wie 1+1=3-Effekte in der Produktion maximiert werden konnen. Z Wirtsch Fabrikbetr 104(5):385–391

Sealy MP, Madireddy G, Williams RE, Rao P (2018) Hybrid processes in additive manufacturing. J Manuf Sci Eng 140(060801):1–13

Thomas CL (1996) Automating sheet-based fabrication: the conveyed-adherent process. SFF Symp Proc:281–290

Varotsis AB, Friel RJ, Harris RA, Engstrom DS (2018) Ultrasonic additive manufacturing as a form-then-bond process for embedding electronic circuitry into a metal matrix. J Manuf Process 32:664–675

White D (2003) Ultrasonic object consolidation. U.S. Patent No. 6,519,500. Washington, DC

Wimpenny DI, Bryden B, Pashby IR (2003) Rapid laminated tooling. J Mater Process Technol 138(1–3):214–218

Yan X, Gu P (1996) A review of rapid prototyping technologies and systems. Comput Aided Des 28(4):307–318

Yang Y, Janaki Ram GD, Stucker BE (2009) Bond formation and fiber embedment during ultrasonic consolidation. J Mater Process Technol 209(10):4915–4924

Zhang Y, Wang J (2017) Fabrication of functionally graded porous polymer structures using thermal bonding lamination techniques. Procedia Manuf 10:866–875
Zhang G, Chen H, Yang S et al (2018) Frozen slurry-based laminated object manufacturing to fabricate porous ceramic with oriented lamellar structure. J Eur Ceram Soc 38:4014–4019
Zhao Z, Yang X, Li S, Li D (2019) Interfacial bonding features of friction stir additive manufactured build for 2195-T8 aluminium-lithium alloy. J Manuf Process 38:396–410

Chapter 12
Future Additive Manufacturing Processes

Abstract Existing additive manufacturing (AM) processes need to be developed in order to demonstrate their potential for product fabrication. There is some scope for development of powder bed fusion by incorporating extra deposition tracks and extra powder hoppers which is given. Incorporating extra nozzle in laser solid deposition process will help make complex products – this is explained. Existing AM processes are not capable to utilize all available resources for product development, classification of AM processes suggests the development of new AM processes that will utilize untapped resources in new way. These new processes could be particle bed process, non-photopolymer bed process and sheet bed process which are given.

Keywords Multimaterial · Multifunctional · Functionally graded · Recycling · Product development · Sheet material

12.1 Introduction

Powder bed fusion does not make multi-material products; if a process is going to be developed which deals with a number of materials simultaneously then it will herald a new beginning in new types of product development. Beam based solid deposition process can be well-known for making large components but it does not match to powder bed fusion in making complex products; if the process will be developed which could provide support from feedstock (powder) then it will be able to fabricate not only a large component but also a complex component. Sheet based process cannot make any product without taking help from machining, what if sheet based process is changed to a sheet bed process which does not do machining.

All feedstocks require careful fabrication because feedstocks need to satisfy certain criteria – if powder is a feedstock then it has to fulfil some standard of size distribution and shape (Spierings et al. 2015), if filament is a feedstock then it has to be of certain thickness and uniformity etc. What if a process is developed which can use a feedstock having no such standards, some material that is discarded and needs to be recycled. This chapter suggests some processes and tries to answer some of these questions.

S. Kumar, *Additive Manufacturing Processes*,
https://doi.org/10.1007/978-3-030-45089-2_12

12.2 Future Processes Based on Classification

Classification given in Chap. 2 provides pictorial summary of AM processes; there are some categories which have abundance of varieties while for some other categories there are few varieties to come under. There are fewer varieties for material bed process while there are many varieties that come under material deposition process. Though, classification in itself does not inform, why there are fewer varieties for some categories, or the reasons for difficulties in processing, or economics about the development of a process variant; but, the classification provides a glimpse to check if there is anything missing. It does not imply that processes which are missing will certainly be feasible and viable. The presence of classification just provides an advantage, which does not remain available in absence of such classification. The advantage is to mark and observe any asymmetry in the development or flow of classification. If there is any asymmetry, then it implies that some category of processes is either under-developed or over-developed – which can lead to find some new processes. The advantage is thus to know some missing processes from the classification. The claim of these missing processes is that they were unknown but due to the classification. Being a missing process does not mean that it is the best yet-to-be-invented process, being a missing process only means that it is yet to be checked whether it can be one of the best processes. Some of these processes are particle bed process, non-photopolymer bed process and sheet bed process that are given below.

12.2.1 Particle Bed Process

Material bed process is mainly of three categories – powder bed process, slurry bed process and photopolymer bed process. Powder bed process (PBP) is the only bed process that is using solid feedstock (as given in Chaps. 3, 4 and 5). PBP provides high complexity that is unattainable by any other AM process for high melting point materials. But, this complexity comes at a cost; powder needs to flow on the bed (Snow et al. 2019); a well-flowing powder is expensive while any powder is not expensive (Rogalsky et al. 2018). Being expensive is not a drawback if a complex part is getting formed. But, what if forming a complex part is not an aim, high surface finish is not desired and porosity is not detrimental; using PBP even then means underutilization of PBP systems and squandering of expensive resources. These non-excellent and non-complex parts, which are many, do not require expensive powders and sophisticated systems; they do not need PBP to be formed. They do not need PBP not because their requirements are high but because their requirements are low.

There are no processes that can cater to those needs; there is actually no process that wants to aim such low to cater to those needs; using solid materials in the context of a bed, there are actually no processes available other than PBP. For fulfilling those needs, a well-flowing powder is not required; these powders are approximately of size range from 20 to 150 micron; for such low requirements, such size is still small, bigger size from 500 micron to 1 mm can work. A powder is more expensive than any non-uniform solid material of equivalent size such as a small wire bit, finely cut machining chips, broken pieces, small flakes, discarded material after grinding and drilling, waste material during powder formation, remnant after casting etc.; if these materials are not of equivalent size then breaking them into smaller size of any shape is easier than producing powders of equivalent size.

There is a process missing which will utilize these discarded materials or particles in the form of a bed and make parts, the process whose primary purpose will be recycling than competing with other AM processes for making sophisticated parts. There is no dedicated AM process available for recycling. Though, all AM processes can claim to contribute to sustainability (Ford and Despeisse 2016), there is no AM process which is exclusively designed to contribute to sustainability, this process will thus fulfil the gap. The process will face less obstacles in its development; the process would have faced more technical obstacles if there were no PBP; the process has advantage to check whether know-how gained from PBP can be applied in its context.

12.2.2 Non-photopolymer Bed Process

Classification shows that there are only two types of liquid bed process, that is photopolymer bed process (PPBP) and slurry bed process; there are a number of liquid based processes of material deposition type already available; it implies that there are many types of liquid workable in AM. Using other liquid types, processes could be polymer bed process, hydrogel bed process, ink bed process, water bed process etc.; these bed processes will create complex structures without having compulsion to make necessary support structures and thus will herald a new area of complex part development using liquid other than photopolymers and slurries.

12.2.3 Sheet Bed Process

AM processes are classified in two group – material bed process and material deposition process. Material bed process is further classified, a modified version is given in Fig. 12.1. Solid bed process is found of just one type, that is powder bed process; no other feedstock type bed process such as wire bed process, sheet, block etc. are found to exist. Observing the classification, these types of solid bed process made up with wire, sheet etc. are the missing link in the classification. Predicting the future using missing link is not new. In case of sheet based process (such as UC, LOM, FSAM as given in Chap. 11), it can be described as a combination of sheet bed plus machining, since machining is involved and these processes could not find

Fig. 12.1 Classification of material bed process: future process, sheet bed in green dotted box

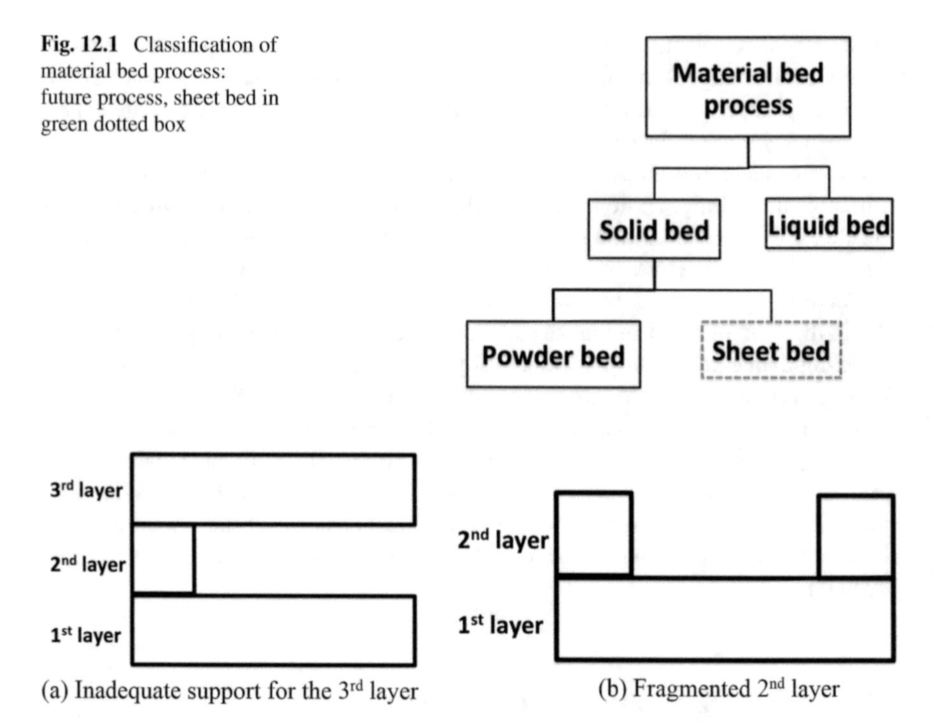

(a) Inadequate support for the 3rd layer (b) Fragmented 2nd layer

Fig. 12.2 Geometries not suitable to be made by SBP: (**a**) due to inadequate support provided by the second layer to the third layer, (**b**) due to lack of connection in left and right parts of the second layer

place in the classification, but it gives a direction that future process can use sheet bed types process where there would not be such machining as found in the present sheet based process (SBP). If sheets are pre-machined as per the requirement of a particular cross-section of a part, then only thing remained is to join these pre-machined and shaped materials. Figure 12.2b shows side view of a solid block to be fabricated or assembled from pre-machined plates as shown in Fig. 12.2a. Since plates A, B and C are already machined, they need to be joined in sequence as shown in Fig. 12.2b. The problem of making parts lie in selecting the pre-machined plates, placing them in an order and joining them; this is in principle not different from powder bed process (PBP) process where powders are placed and joined. The difference is that in PBP same types of powders are placed all time while joining creates a difference and different parts are formed; in SBP different types and design of sheet metals are planned to be placed while joining is always same; this different placing will create a difference and will be responsible for making different parts.

Figure 12.2 shows a simple part made by three plates; these plates can be placed and joined by any techniques appropriate to the material and size, it could be adhesive, glue, welding or these could be joined by diffusion sintering during post-processing (Roosen 2001). But what is going to happen if complexity and the size increases – it may require hundreds of pre-machined plates of various cross-sections as per the design and then a method is required which will place them in a sequence and bond them. If design and size change, hundreds of more plates are required and so on. Thus, ten different parts are fabricated with the help of more than thousand plates; these plates are cut as per the design of the part; thus cutting follows the design of the part, for making many parts cutting has to follow the design many times. This method of manufacturing can certainly be a new additive manufacturing but it is certainly not a rapid manufacturing – it goes so slow, it waits for the design to come, it then waits for the machine to start and cut the plates. If it is not rapid manufacturing, then even though it is additive manufacturing it might not be technically interesting. The solution lies in finding an alternative – if design follows the cutting instead of letting the cutting to follow the design; a thousand of parts could be fabricated by less than a thousand of types of plates. If thousands of pre-machined plates of various designs and sizes are kept in a workshop, these plates have potential to make thousands of parts; these plates are waiting to be picked up by a robot hand and kept somewhere in a sequence to be assembled and integrated, or these plates are waiting to be picked up by a CNC machine in the same way as a CNC machine uses to choose different tools for different types of machining. If a part is to be fabricated, computer identifies appropriate plates to be selected; these are then physically selected and then joined to make a part; this methodology (computer aided process planning) (Abdulhameed et al. 2018) is not new but has not yet been applied in sheet based AM. In this way, thousands of designs can be fabricated by pre-cut plates, design follows the cutting. Thousands of designs can be fabricated but not all designs can be fabricated as pre-cut plates need to be available to match the design. This brings limitation to this process while other AM processes have no such limitation, they can make in principle a part of any design. This limitation is in addition to the common limitation of the SBP related to the geometry. If there is no adequate support for the successive sheet (Fig. 12.2a) or there is no connection between two parts of a sheet (Fig. 12.2b) then these types of geometry are not suitable to be tried by any SBP unless adequate measure is taken.

Advantages of this process are following:

1. There is no defect or porosity in a pre-machined plate or in a layer due to the process; the layer is acquired and is not built by the process therefore the process-induced defects remain absent; this advantage is common in many other sheet based processes. In other AM processes, where a layer needs to be made from powders, wire, photopolymer liquid, polymer paste, slurry, polymer liquid etc., the layer formation brings defects such as porosities, inhomogeneities, inter-line bonding defects, side roughness, top roughness etc.; these defects are avoided in this process unless some of these defects are intentionally introduced in pre-machined layers in order to fulfil the requirement of a part. These defects, which are both avoidable and unavoidable depending upon the geometries, induce limitations in surface roughness and accuracy while this process will not be having such limitations. This gives rise to a question – which type of limitation is more severe – a limitation in the geometry due to unavailability of pre-machined lay-

ers or a limitation in a geometry due to inability to make layers, while the former limitation is due to logistics the latter limitation is due to the process itself, while the former limitation can be overcome without a need to change the process the latter limitation cannot be overcome without overcoming the limitation of the process.

2. The process promises to fabricate overhangs and undercuts with ease, these features can be incorporated in a part during machining stage, that is pre-joining stage. As shown in Fig. 12.3, an overhang which is an upper part of a circular hole is introduced in a part (Fig. 12.3b) by making a hole in plate B and then by joining with plate A and plate C. For making such holes or such overhangs in an AM process, either supporting structures are required or feedstock becomes the supporting structure – due to this it is not possible to make overhangs in all geometries, and even when overhangs are made their surface quality is of no match to that made by machining. AM processes find it more difficult to make overhang structures or features than to make non-overhang structures or features. The more difficulty they face, the less complexity they can offer. It is rare to have a very complex part having no overhangs. Overhangs can be elaborated as slanted walls, slanted pins, slanted pillars, tapered hole, side hole on a vertical wall, part of an undercut, side hole on a pin etc. There is no AM process which claims to be able to fabricate a complex part and not be able to include such overhang features in the part. This process promises to offer solution not because it has found solution how to join bit by bit like other AM processes join but because it does not try to join bit by bit, since this process does not try that way this process does not have possibility to fail that way. The problem of making overhangs by joining is converted into a problem of largely machining in this process. As circular hole (Fig. 12.3) was created by simply drilling a plate rather than by making a number of thin layers from small pieces and optimizing the process in order to be able to join those layers. Drilling gives higher surface finish, accuracy and homogeneity than any AM process. If circular hole is big so that it will require more than one pre-machined plates then the incorporation of such hole in the part will not only governed by machining but also by assembling and joining, but joining in this case is different than the joining in other AM processes because this joining will still not be able to disturb the accuracy and surface finish of an individual semi-circular part of the hole which solely belongs to an individual plate. If in-process joining gives inaccuracy, the process gives another option to accomplish the joining during post-processing stage; an in-

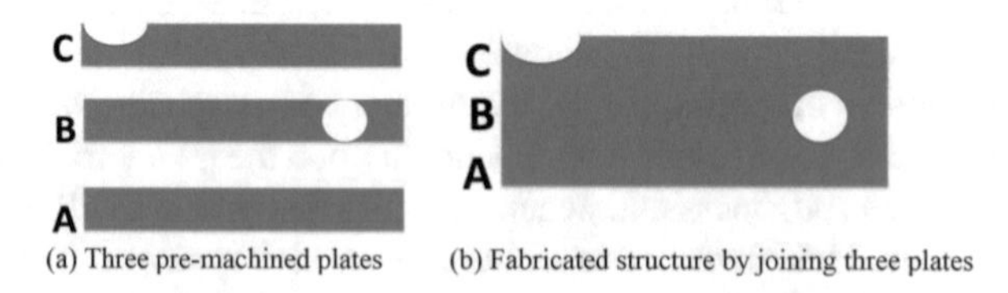

Fig. 12.3 Side view of pre-machined plates and a structure fabricated by joining these plates: (**a**) three pre-machined plates named A, B and C, (**b**) fabricated structure by joining these plates

process joining such as welding (in case of sheet metals) which tends to give inaccuracy due to excessive melting can be replaced by post-process diffusion sintering of an assembled pre-product. In those examples where welding is replaced by post-process sintering, the strength achieved will not be the same but it is better to have a complex part possessing no such high strength than to have a no-part possessing high strength.

This process can be especially useful for making parts which are at the two ends of a scale of complexity and size –a very big part (one dimension of the size of a metre) having low complexity and a very small part (one dimension of the size of centimetre) having high complexity. An example of a big part of low complexity is a metallic tool for automotive applications while an example of a small part of high complexity is a metallic gear or a part equipped with sensor or electronic circuits.

12.3 Future Powder Bed Process

12.3.1 Using Powder Deposition Tracks

All PBP systems employ single powder deposition mechanism as shown in Fig. 12.4a; powder is deposited on a build platform by carrying powder in AB line on the same plane. It facilitates fabrication of a part made due to contribution from single type of powders either it could be a specific metal, ceramic, polymer or composite. This restricts fabrication of a multi-material part, fabrication could have been possible if there were additional powder deposition facilities. Figure 12.4b demonstrates a possibility to develop multi-material deposition systems in a powder

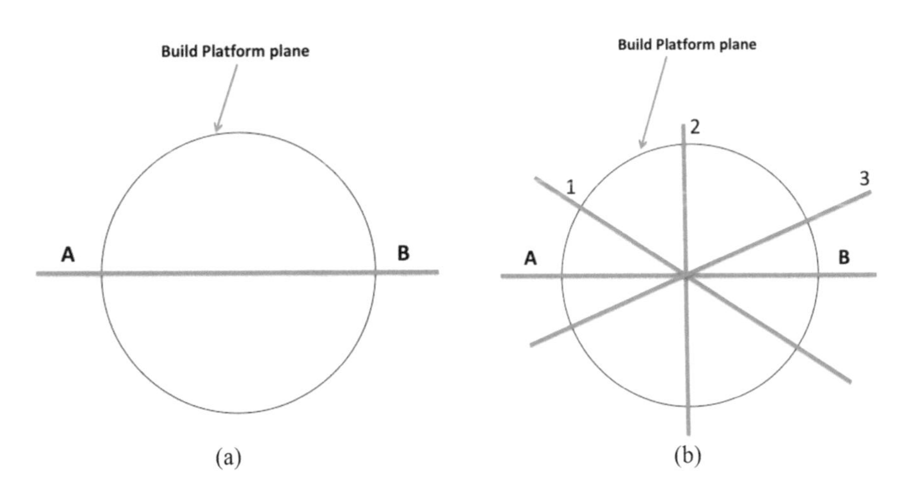

Fig. 12.4 Powder deposition movement in (**a**) PBP system deposition track AB (**b**) possible modification using powder deposition tracks 1, 2, 3 and AB

bed fusion (PBF) machine; straight lines 1, 2, 3 show additional tracks other than original track AB for depositing powder on the same build platform facilitating deposition of four types of powders. Figure 12.5 shows a schematic diagram for powder deposition using three tracks. Extending it for multiple tracks will increase the dimension and the complexity of the PBP system; employing circular tracks instead of straight line tracks is an option to optimize the dimension. There are following main challenges to implement the modified process:

1. *Contamination of powders*: After solidification of the deposited layer and before laying another layer using different powder, platform and chamber need to be cleaned to ensure there is no mixing of two types of powders; mixing will cause a decrease in quality of recycled powders.
2. *Delay in fabrication*: PBP works in recurring steps of deposition of materials and selective solidification; henceforth it needs to deal with an additional step of cleaning and collecting the powder. This will increase the fabrication time and make AM slower.
3. *Increase in machine dimension*: For each increase in deposition track, two containers one for storing powder and other for overflowing powder (as shown in Fig. 12.5) are required, which will increase the dimension of the machine, could be compensated with a decrease in processing chamber, resulting in a smaller part.
4. *Selection of materials*: All materials belonging to different powders need to be metallurgical compatible with each other so that they do not make deleterious compounds at the interface during solidification. Selection of materials should also be done keeping in view their effect on recycling when they get inadvertently mixed during processing. For instance, mixing of SS 304 and SS 316 will not cause the same deterioration in the properties of the part made from their recycling than the mixing of SS 304 and alumina will. Two materials with high difference in melting point and properties (reflectivity, conductivity) will additionally pose challenges in optimizing process parameters.

Following advantages are envisaged from the modified process.

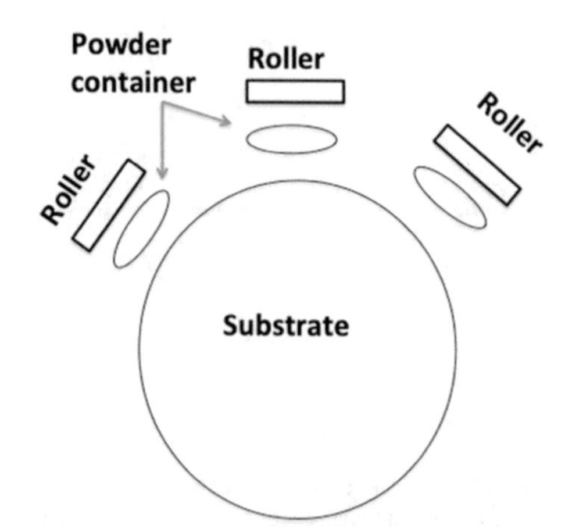

Fig. 12.5 Schematic diagram of top view for powder bed system using three powder deposition tracks

12.3.1.1 Fabrication of Functionally Graded Materials

There are various techniques by which functionally graded materials (FGM) can be made in the present PBP systems (Loh et al. 2018) – (a) by creating various degrees of porosities resulting in graded properties, (b) by creating a porous part and infiltrating with it various infiltrants and (c) by replacing the powder at various intervals during processing. All these techniques have limitations; in technique (a) there is one material which limits the range of properties achieved, in technique (b) infiltration limits the size of a part and the melting point of materials to be used, in technique (c) material replacement requires interruption of on-going build which compromises the quality of a part and brings enormous delay in fabrication time.

The modified process will have provision to use a number of materials and will be able to make FGM; Fig. 12.6a shows an example of FGM made from three materials A, B and C. After using material A, system will be cleaned with material A without opening the chamber. Material B will then be used followed by cleaning the material B and using the material C resulting in FGM part.

12.3.1.2 Fabrication of Multi-functional Parts

The modified process will facilitate fabrication of a part having different materials in the same layer resulting in a multi-material multi-functional part having each material planned to serve specific function. Figure 12.6b shows an example of multi-material part comprised of three features A, B and C made from three materials A, B and C respectively.

Figure 12.7 shows various steps for fabricating the part shown in Fig. 12.6b. In the first step as shown in Fig. 12.7a, feature A of the part is made by using four layers of material A. In the second step as shown in Fig. 12.7b, a gap equal to layer thickness of material B is created – a known step in any PBP. In the third step (Fig. 12.7c), material B is used to make feature B. In order to make feature C, material B is cleaned and material C is used as shown in Fig. 12.7d. After third step, no gap as done in second step is created as feature C needs to be fitted in the same layer.

In this way, a multi-material part comprising of four layers of A and single layer of B and C is fabricated. With an increase in height of feature B and C, making

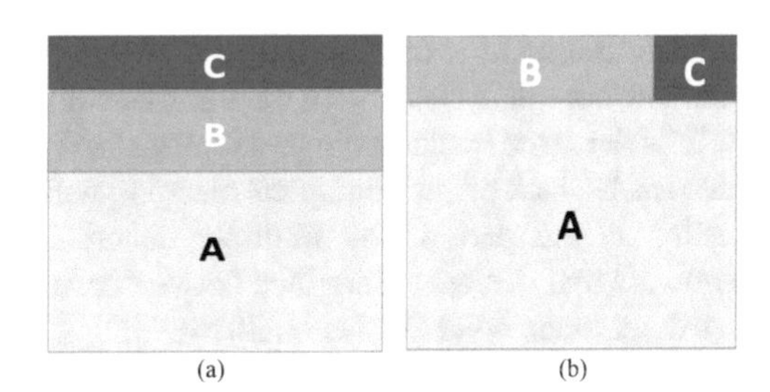

Fig. 12.6 Schematic diagram of (**a**) functionally graded material deposition, (**b**) multi-functional part

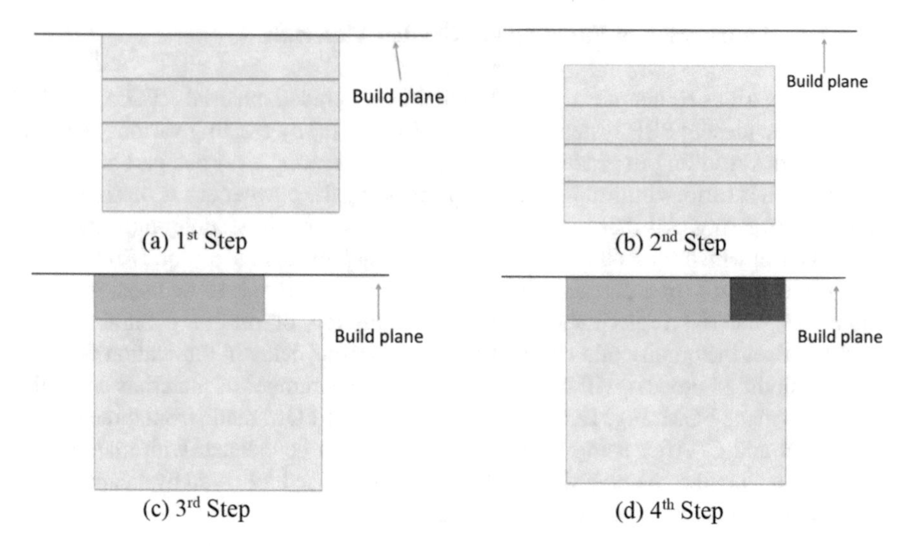

Fig. 12.7 Various steps of fabrication of a multi-functional part: (**a**) 1st step, (**b**) 2nd step, (**c**) 3rd step, (**d**) 4th step

feature C (fourth step) will not be easy as it will require to be solidified in one step; for feature C consisting of two or three layers optimizing the experimental parameters will be relatively easy than that for consisting of several layers. This shows the limitation of modified process.

12.3.2 *Using Powder Hoppers*

12.3.2.1 With Roller and Cleaning

Instead of using a powder bed fusion system consisting of a number of powder deposition tracks (Fig. 12.4b, multi-material parts can also be fabricated by an alternative method using a number of powder hoppers containing different powders A, B and C as shown in Fig. 12.8. In order to make a multi-material part, either powder A, B or C will be used to make certain number of layers. After cleaning the chamber, another powder will be used to make certain number of other layers until all layers are completed. Number of hoppers can be increased or decreased to incorporate desired number of materials in a part. In this method compared to the previous method, complete cleaning of the track is required as different types of powders will follow the same track; in the previous method where separate powders follow separate tracks, lack of such complete cleaning will not furnish same degree of contamination in the part. Using multiple materials will require development of new software and design of interface between two materials so that there will not be a weak interface bond (Wei et al. 2018).

12.3.2.2 Without Roller and Cleaning

In this case, powders are not levelled and therefore no roller or scraper is used; besides, cleaning as an intermediate step as done in the previous case is not required and therefore not employed. Hopper is used as a source of powders but it can also be used as an apparatus to deposit powder in a line and create a deposition track. If there are several hoppers having different powders, they can be used for depositing a number of lines of different powders in a plane giving rise to a multi-powder or multi-material layer – a start for creating a three-dimensional multi-material structure. Replacing several powder hoppers with three or four powder hoppers and an array of micro-hoppers will still facilitate to make a multi-material structure using limited number (three or four) of materials (Das and Santosa 2001). Powders from bigger hoppers can be mixed in a number of combinations and the resulting mixtures will be deposited through an array of micro-hoppers to make lines (Fig. 12.9). Controlling the flow will determine the height of powder lines; consequently, a layer having various powder lines of equal height made from different powders can be deposited implying there is no need of roller for layer height control. Though, height of the line can be controlled, but flow control will not help control the geometry of the line which is determined by inter-particle friction (angle of repose) and is different for different powders.

Figure 12.10 shows front view of deposited powder lines; for no overlap between two adjacent powder lines there is a large gap (Fig. 12.10a, c) which will become

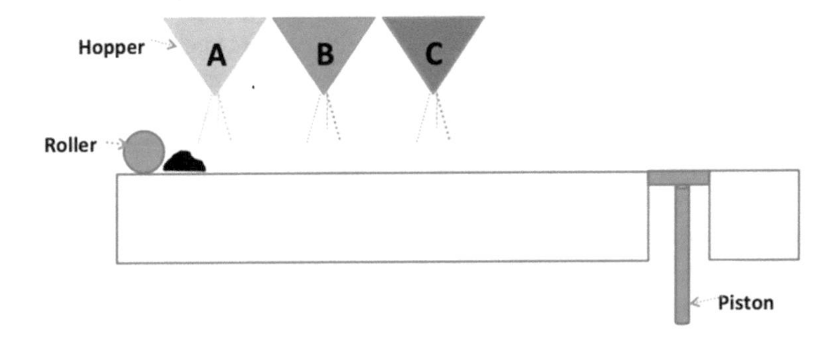

Fig. 12.8 Schematic diagram for making multi-material part using three powder hoppers containing powders A, B and C respectively

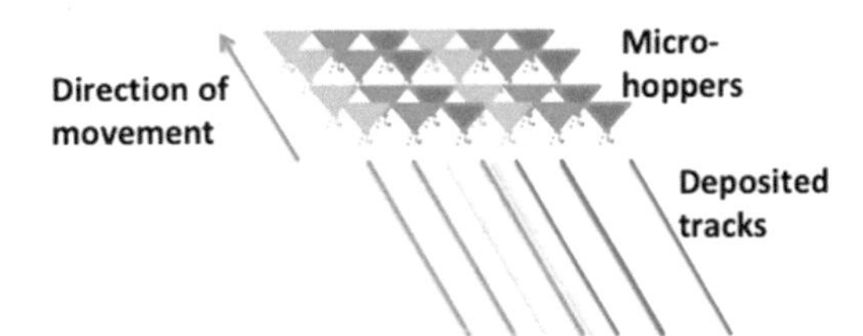

Fig. 12.9 Powder tracks in the form of lines are deposited by movement of micro-hoppers

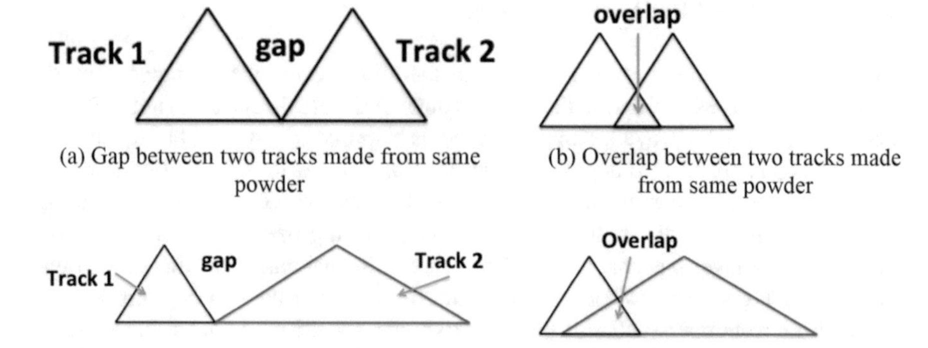

(a) Gap between two tracks made from same powder

(b) Overlap between two tracks made from same powder

(c) Two adjacent tracks made from different powders

(d) Aymmetric overlap between two tracks made from different powders

Fig. 12.10 Front view of two powder lines deposited by micro-hopper: (**a**), (**b**) using same powder; (**c**), (**d**) using two different powders

source of porosity and defects when powders will be processed to form a solidified layer. These gaps can be reduced by overlapping two tracks (Fig. 12.10b, d) which in case of different materials will give rise to an asymmetric overlap (Fig. 12.10d) – a source of an optimization problem for placing right amount of materials and still having no defects, the problem will increase with an increase in the number of layers and types of materials. Though, geometry of the overlap will depend upon the time and sequence of deposition and may vary from Fig. 12.10b, there will still be a marked difference between overlaps made from similar and different powders. This brings to the fore that using roller has distinct advantages in powder deposition – the roller does not create different geometries during the deposition, it is not meant for that purpose – and therefore the problems associated with creating geometries during deposition stage are not found when a roller is used; geometries are created during fusion stage and therefore the problem is only at the fusion stage rather than at both stages. The previous processes which employ rollers are better for making multi-material integrated parts. But, for certain geometries when lines are distinct (such as scaffold), there is minimum interaction between two different materials (outer wall is made from insulating material while inner wall is made from conducting material), creating features on non-planar surface etc.; the present process will have an advantage.

12.4 Future Laser Solid Deposition Process

Laser solid deposition process (LSDP) is not suitable for making overhangs as a powder stream coming from a nozzle is unable to create supporting bed for further deposition on it. Though, there are other ways to overcome this requirement and make overhangs. Figure 12.11 shows a side view of 3D structure BACDEFG having

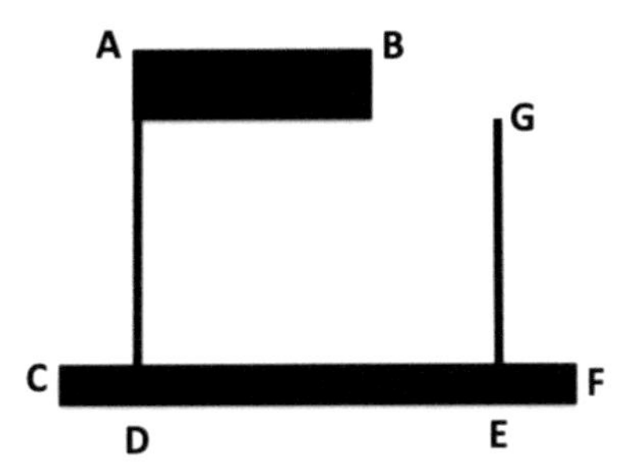

Fig. 12.11 A 3D structure BACDEFG having an overhang AB, two pillars AD, GE and a base CF

an overhang AB, two walls AD, GE and a base CF requires to be fabricated to evaluate the capability of LSDP. The process can make two walls and base (as shown in Fig. 12.12c) as for making them no supporting material is required but the process cannot make overhang AB. Changing the orientation of this design will not help as in any orientation there will still be some overhangs to be built.

If the process is going to be executed in a setup equipped with a CNC machine having capability to change the orientation of the substrate (fixed on a platform), then it will be possible to make same overhang AB. Fig. 12.12 shows various steps for making overhangs: in step 1 (Fig. 12.12a), basic structures that could be made by LSDP is made; in step 2 (Fig. 12.12b), the orientation was changed by changing the position of the platform; in step 3 (Fig. 12.12c), overhang AB was made, in this step overhang is no longer an overhang but just like a wall AB supported on another base AD (former wall AD); step 4 (Fig. 12.12d) shows a fabricated structure similar to a desired structure (Fig. 12.11) still attached on the substrate which needs to be cut off and separated.

Though in this example, it is possible to fabricate a complete 3D structure just by taking resort to orientation change but there are many examples with a higher degree of complexity where such manoeuvring will not help. For example, if in 3D structure shown in Fig. 12.11, height of the wall will be increased, then the laser beam or powder stream coming from the nozzle will not be able to reach AD and thus fabrication of overhang will not be possible. If wall AD is weak, then it will not be able to bear the weight of AB and will collapse during the processing. If the separation between two walls AD and GE is large, then the nozzle will not be able to reach near AD without it or its accessory being collided with GE. Improving the process by modifying the system can enhance thus its ability for fabrication. Figure 12.13 shows possible modification in which an existing nozzle is fitted with an extra nozzle to deliver the powder and filling the gap so that overhangs can be fabricated over filled-up powders; these powders will sum up as a supporting plane somewhat similar to a powder bed.

Figure 12.14 shows various steps for fabricating the structure (given in Fig. 12.11): in step 1 (Fig. 12.14a), two walls and base are made; in step 2 (Fig. 12.14b), the gap created by two walls is filled-up using powder fed by an extra nozzle, surface smoothness of filled-up powders can be improved by optimizing parameters such as filling rate but it will not be as smooth as created by a roller in PBF, though density of filled-up powders in both processes will be same as powders

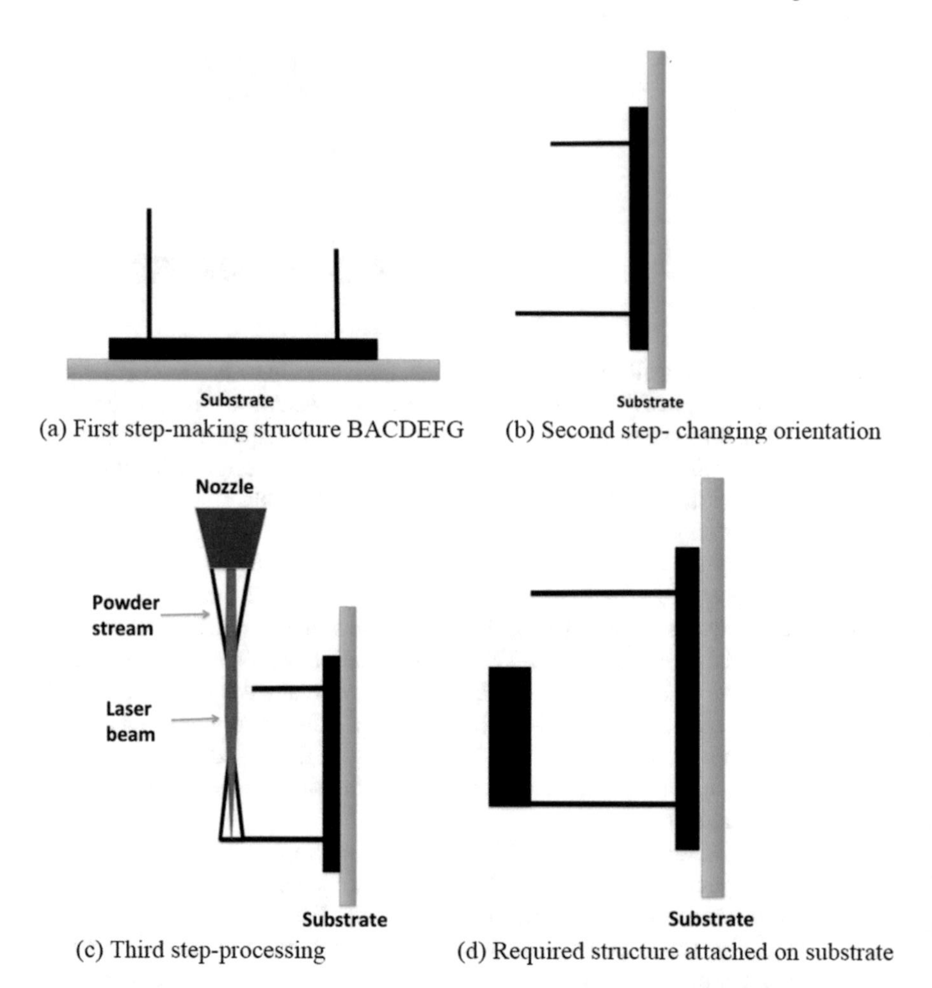

Fig. 12.12 Fabrication of structure (shown in Fig. 12.11): (**a**) first step – making basic structure by LSDP, (**b**) second step – changing the orientation by rotating the platform by 90°, (**c**) third step – building the overhang, (**d**) required structure attached on a substrate to be cut off

are not pressed in PBF; in step 3 (Fig. 12.14c), overhang is made over these powders; for making overhang or filling the gaps powders used are same so that the possibility for recyclability of powders will not decrease; in step 4 (Fig. 12.14d), powders are drained out and the structure is separated from the substrate. If the orientation is turned upside down and the structure is fabricated, powders will not be drained out until the structure is separated from the substrate. In this case, the height of both walls (Fig. 12.14a) is made same while in previous case (Fig. 12.12a) the height of one wall AD is extended equal to the height of the overhang so that the overhang will be built on the extended part of AD. Consequently, final structures fabricated in both cases will not have the same properties and microstructures.

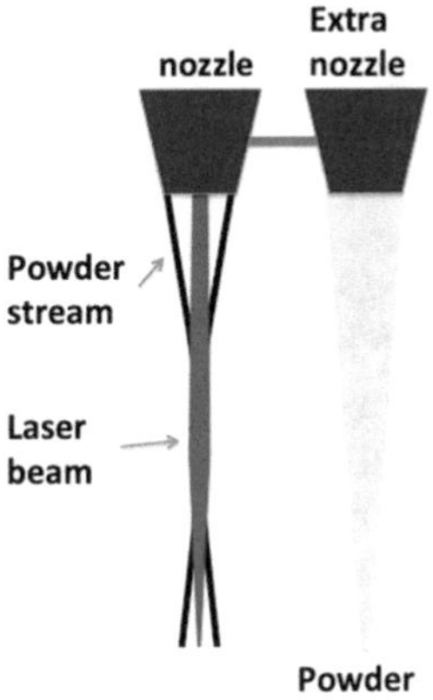

Fig. 12.13 Modification of an existing nozzle by attaching with it an extra nozzle for delivering filling powders

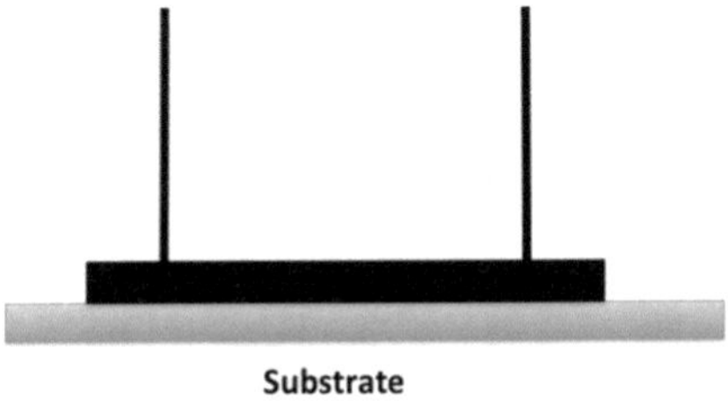

(a) 1st step- fabrication of basic structure by LSDP

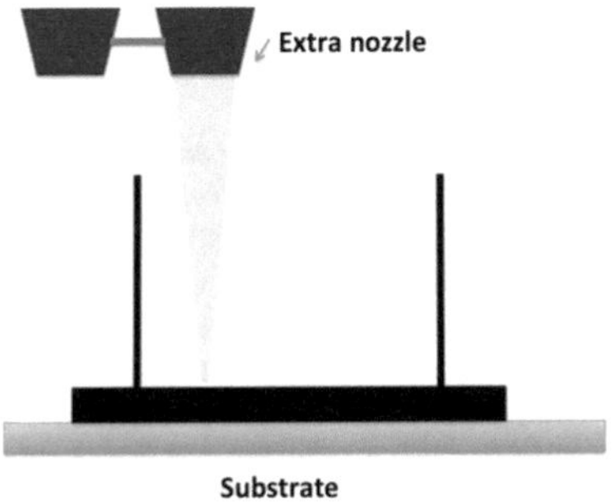

(b) 2nd step- filling up the gap using extra nozzle

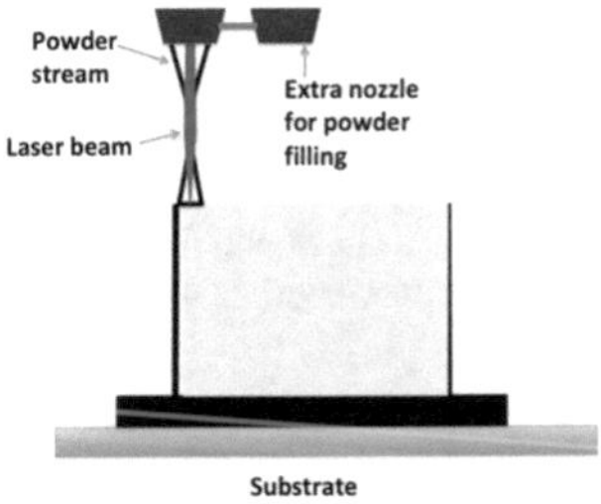

(c) 3rd step: fabrication of overhang over filled-up powders

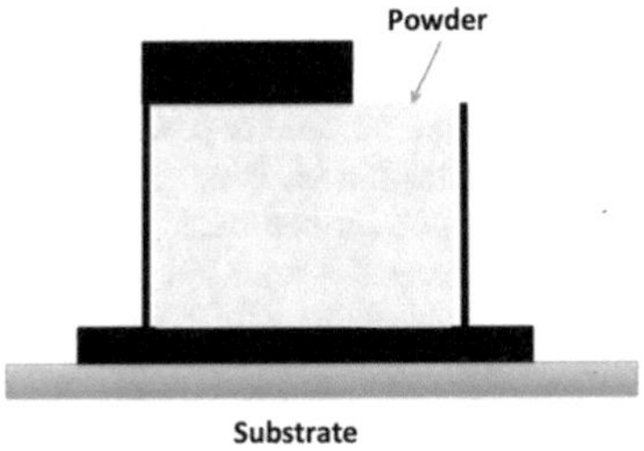

(d) 4th step: powders to be drained out and structure to be cut off

Fig. 12.14 Fabrication of structure (shown in Fig. 12.11) using a modified nozzle in four steps: (**a**) 1st step, (**b**) 2nd step, (**c**) 3rd step, (**d**) 4th step

This brings to the fore that the same structure fabricated by two variants of the same process will result in variation in properties though the intention was not to vary the properties while making the same structure. This is where AM sharply differs from machining; in machining irrespective of the sequence of operation in a conventional machine (CNC machine, lathe, milling, drilling etc.), the material properties will not be different in final structure. This does not mean that machining has no implication in material properties – machining may or may not have implication in material properties but sequence of operation in machining certainly has no implication in material properties, if machining starts from top side of the block and ends in the bottom side or vice versa, machining will furnish same properties. The same cannot be true in AM, if the sequence of operation is starting from making a top layer or top feature of a part and ending in making bottom layer or bottom feature of the part, and vice versa; in both sequences of operation, material properties will be different. Consequently, if the sequence of operation is lost, a complex part fabricated again in an AM machine will not give as much repeatability as a complex part fabricated again in a CNC machine gives. Other differences in machining and AM is given in Chap. 1.

References

Abdulhameed O, Al-Ahmari A, Ameen W, Mian S (2018) Novel dynamic CAPP system for hybrid additive-subtractive-inspection process. Rapid Prototyp J 24(6):988–1002

Das S, Santosa J (2001) Design of a micro-hopper array for multi-material powder deposition. SFF Symp Pro:155–162

Ford S, Despeisse M (2016) Additive manufacturing and sustainability: an exploratory study of the advantages and challenges. J Clean Prod 137:1573–1587

Loh GH, Pei E, Harrison D, Monzón MD (2018) An overview of functionally graded additive manufacturing. Addit Manuf 23:34–44

Rogalsky A, Rishmawi I, Brock L, Vlasea M (2018) Low cost irregular feed stock for laser powder bed fusion. J Manuf Process 35:446–456

Roosen A (2001) New lamination technique to join ceramic green tapes for the manufacturing of multilayer devices. J Eur Ceram Soc 21(10–11):1993–1996

Snow Z, Martukanitz R, Joshi S (2019) On the development of powder spreadability metrics and feedstock requirements for powder bed fusion additive manufacturing. Addit Manuf 28:78–86

Spierings AB, Voegtlin M, Bauer T, Wegener K (2015) Powder flowability characterization methodology for powder-bed-based metal additive manufacturing. Progr Addit Manuf:1–12

Wei C, Li L, Zhang X, Chueh YH (2018) 3D printing of multiple metallic materials via modified selective laser melting. CIRP Ann Manuf Technol 67:245–248

Index

S. Kumar, *Additive Manufacturing Processes*,
https://doi.org/10.1007/978-3-030-45089-2

Printed in the United States
by Baker & Taylor Publisher Services